How to Grow Vegetables

How to Create a Thriving Vegetable Garden

(Step by Step on How to Grow Seeds Organic Vegetable Seeds Healthy Diet Plan)

Romeo Willard

Published By **Chris David**

Romeo Willard

How to Grow Vegetables: How to Create a Thriving Vegetable Garden (Step by Step on How to Grow Seeds Organic Vegetable Seeds Healthy Diet Plan)

ISBN 978-1-7771462-2-1

Legal & Disclaimer

The information contained in this book is not designed to replace or take the place of any form of medicine or professional medical advice. The information in this book has been provided for educational & entertainment purposes only.

The information contained in this book has been compiled from sources deemed reliable, and it is accurate to the best of the Author's knowledge; however, the Author cannot guarantee its accuracy and validity and cannot be held liable for any errors or omissions. Changes are periodically made to this book. You must consult your doctor or get professional medical advice before using any of the suggested remedies, techniques, or information in this book.

Upon using the information contained in this book, you agree to hold harmless the Author from and against any damages, costs, and expenses, including any legal fees potentially resulting from the application of any of the information provided by this guide. This disclaimer applies to any damages or injury caused by the use and application, whether directly or indirectly, of any advice or information presented, whether for breach of contract, tort, negligence, personal injury, criminal intent, or under any other cause of action.

You agree to accept all risks of using the information presented inside this book. You need to consult a professional medical practitioner in order to ensure you are both able and healthy enough to participate in this program.

Table Of Contents

Chapter 1: The Basics Of Container Gardening

Choosing Containers

If you are choosing the right container to plant in, there are a variety of factors to consider. Included are the types of materials, colors and sizes and the shapes.

Did you know there are numerous decisions to make when selecting containers to plant your gardening? If you are deciding which pot you will need, think about how often you will need to water and the weight of the pot as well as the cost.

One of the first things to think about in deciding on containers to use to plant in is the materials the container is constructed of. There are a few pluses and negatives you should consider when making your choice:

Wooden Containers

Most containers are built of wood. They include window boxes as well as the ever-popular semi whiskey barrels.

Pots of pressed fiber (made out of wood fibres) are thought to be semi-porous. That means that you don't need to water your garden pots more often than you do with the terra clay (porous) However, much more frequently than using plastic pots (non-porous). One of the main disadvantages of wooden containers is that it is susceptible to decay. One method for overcoming this is to line the wooden planter with plastic in order to stop the procedure. There is also the option of having lines of metal made by the machine

If you are planning for using the wooden containers for many years in the future. If you choose to purchase wooden containers, ensure that you purchase resistant to rot, such as redwood or cedar; if not you'll be replacing your containers frequently.

One thing to remember when you are using wooden containers is that a lot of them are made with wood preservation agents. When you grow vegetables inside containers, you must be 100% certain that the containers been used to store products that are unlikely to cause harm to the either the plants or to people. Wood used around plants shouldn't be treated with creosote, or wood preservation agents. They could be poisonous for plants, and potentially harmful for people too. The preservatives could get into the soil when irrigation, and as such that using wooden containers to store veggies, fruits as well as other foods may not be the most beneficial option in case you're concerned about the health of your family members.

Terra Cotta Containers

If you are looking for containers to use in containers for gardening, you can see a wide selection of planters made from terra cotta and containers to choose from. To create a natural appearance that's always good fashion, terra-cotta ranks first in the market. The earthy, orange color can be a perfect complement to the green leaves of the plants.

There are some drawbacks to using terracotta when picking the best container to plant your gardening. Terra cotta isn't frostproof and it is easily damaged and chipped. It is heavy if filled with pot mix, and is difficult to move.

Due to its porous nature and pliable, it requires frequent irrigation to keep the plant and the container hydrated.

Ceramic and Pottery Containers

Pottery is akin to the terracotta material in that it's porous. If you are looking for outdoor pottery, there are a variety of beautiful container designs. It is important to consider when you are choosing the right container to plant in is the long-term durability of outdoor ceramic depends upon the firing method as well as the process of manufacturing. It is difficult to determine the high-quality of the ceramic especially since it's produced in a variety of countries.

The outdoors pottery, specifically clay varieties, has regularly watered similar to the terracotta pots, in order in order to stop the drying process from damaging the plants that are in the pots. When temperatures are cold outdoors, the pottery will require indoor transport in order to stop cracking and freezing of the outdoor pots.

Glass, Plastic and Fiberglass Containers

Plastic, glass and fiberglass containers are not porous products. This means that there is no need to water your containers as frequently like you do with pots that are porous. However it is important to monitor your containers as closely like you would the terra cotta or pottery pots in order to ensure that you do not over-water the plants in your containers. The excess watering of your plants can cause the same damage as letting the plants get dry. If you are constantly watering your pots the plants could be affected by root rot or and various other diseases of plants.

Garden containers made of fiberglass and plastic can also benefit from being extremely light, which means they're excellent choices to hang baskets or window boxes.

Stone Containers

Stone containers are gorgeous but yet heavy and difficult to transport. If you choose to utilize stones when selecting containers for garden containers, they must to be kept on a fixed location due to their difficulty to move about. Concrete and reconstituted stones can be cheaper options.

Stone containers aren't ideal for all kinds of vegetables. The problem with stone for your veggies are concerned is the fact that the stone can heat up fast to temperatures that reach high.

Paper Pulp and Coconut Coir

Paper pulp appears as the texture of a lumpy and bumpy piece of cardboard. coconut fiber containers are characterized by an elongated, shredded appearance. The containers are utilized as liner for windows and wire baskets. They're temporary however, with proper care, they can last up to a period of 2-3 years.

Metal Containers

Copper, brass and lead containers can leak harmful components into the soil. These containers must be avoided while plant vegetables in containers.

Frost Proof Containers

Ceramics that resist frost are accessible in all areas of the outdoor planter. Outdoor pottery comes in various styles, and can also be bought in a variety of colors and glazed designs.

The all-weather garden pots are able to be used throughout the winter. The outdoor pottery for all seasons can stand up to any form of cold winter weather and ease the dry conditions that can be experienced during summer's scorching heat. There are many stores that have designated spaces where they showcase these kinds of containers for garden use. You can spot an attractive banner with a snowflake, which represents the worldwide logo of the frost-free pots. They come with the promise of a refund, so ensure that you save your receipts for the event that your pots fail to perform as they were promised.

Grow Bags

Grow bags are relatively modern accessory to the container garden. They are narrow, long bags constructed of either dark green or black plastic that is reinforced. There are slits that cut through bags and hooks to allow to hang.

The bags are not without controversy. Flowers are awed by these bags however, many feel that they're doing better with their veggies with standard containers. If you are interested take a look at some or two of them for a test of comparison and then decide which one you like best.

Specialty Planters

A lot of gardeners prefer to cultivate their garden in specialty containers like wheelbarrows, strawberry jars and wagons, to create an impact or a statement for their home. There are many of the items on sale in yard sales or even on online sites such as Craig's lists. Be sure to ensure that the container you choose to plant is able to drain at the bottom of the pot.

Some General Guidelines For Choosing Containers

Here are some suggestions and suggestions to bear in mind while choosing your containers for your garden:

The decorative pots that do not have drainage holes will not work to be used for gardening in containers. Four drainage holes at minimum should be drilled into the bottom of every container. Do not be discouraged if you find no holes in the container that you like. Holes can be easily made into containers.

Mix with your containers to create a unique appearance and to add some interest to your garden. Like home décor the garden of your container can be a reflection of a style. It may be classic, Tuscan, contemporary or any of the other types. Take into consideration the design you want to create when choosing the containers for your garden.

The dark colored pots are able to absorb heat, which can cause damage to the roots of plants. If you decide to utilize darker colored containers put them in shaded places will help ensure they stay cooler and will ensure that delicate roots are protected.

Take into consideration the location and accessibility in selecting containers. Select containers that are able to support the capacity if the plant is to move frequently or hang on an eave or a window.

In the case of choosing containers to grow your garden the majority of the time, it is likely to be to enhance the appearance of your garden. We all have our own choices, whether it is wood, terra cotta or even plastic. The main reason we choose these is style as well as cost, and the easy maintenance. If you make a choice, be certain to consider the advantages and disadvantages of different garden containers. Pick the qualities which are most important for you, whether that is aesthetics, cost or maintenance ease. Then decide from there to ensure that you're happy and content with the container garden you have chosen.

Chapter 2: Choosing The Correct Size Container

One of the most important gardening decisions that gardeners need to take is choosing the right container dimension. Gardeners who are new to gardening have no idea of just how big the root structure is that of mature plants. The majority of new gardeners place their veggies in pots which are too small.

Here are some tips to provide an idea of the best size pot for vegetable gardeners.

If a dish is identified as a six" pot, the measure is the measurement of the highest point of the pot. The height will usually be identical to the top of the container.

Here are some useful dimensions for containers of vegetables:

A good depth for one container is 8-12 inches in depth.

The 4-5 gallon containers work for plants that have an upright, bushy appearance like eggplant, tomatoes and peppers.

A few measurements for rectangular pots can be found as follows:

* 24"x36"x8" deep

This size is ideal for root vegetables like beets, carrots, turnips as well as onions.

* 12"x48"x8" deep

This size is ideal for climbing veggies like peas, cucumbers, and pole beans. If you place it against the wall it is possible to create a trellis or install the wire or string to allow this vegetable to expand horizontally.

A great guideline on specific container sizes for every vegetable, refer to Table 10: Recommended Container Sizes.

* Rectangular window boxes are great size for cultivating vegetables. These are especially suitable to grow salad greens. It's a great idea to keep the greens in a window connected to

your kitchen window. Do you want to eat some salad? Go to the window and choose fresh, greens. Add the occasional radishes or some onions!

If you're not sure what size of pot to choose pick the largest one from the options. It is better to choose the larger size pot instead of trying to move an old pot out of the container it was placed in.

Hanging Baskets

Since space and land has been a problem for many people, vertical gardening is now the most popular way of gardening. The most important element of the vertical garden is hanging baskets. A variety of vegetables can be cultivated within hanging containers. The salad greens, like lettuce and others are very effective here. Scallions and cherry tomatoes work very well too.

Upside Down Containers

The most recent development within container gardening is upside-down planters.

These are self-watering plastic bags that come with a hanger hook that is hung from balconies, eaves, and shepherd hooks. Certain of these plants include a reservoir for water which can be filled with water with no disturbance to the plants. Watch a short video of how to plant this topsy-turvy tomato gardener.

Locating Your Vegetable Container Garden

The appeal of gardening in containers is that you are able to make the most of any area you've got. No matter where your home is, you'll locate space to put in a couple of containers. Containers are often placed on the porch, in the stairs or decks. They can be used to mark walkways and paths, as well as be placed along fences to are used to mark boundaries of property. They are able to hang on eaves or climb up trellises.

If you reside in apartments, these plants can be placed in rooftop gardens, on balconies or in window boxes. There is also the option of growing vegetables inside your kitchen or on sunny windows.

This being said, exactly what's needed to produce vegetables using containers.

How Much Sun Do Vegetables Actually Need?

Plants make their own food by an process called photosynthesis. They take energy from the sun's rays and transform it into chemical energy which helps to grow. The more sunlight they get their energy, the more they are able to put into cultivating and generating fruits, foliage and seeds.

A lot of gardeners believe that all veggies require sunshine for optimal development. True, some veggies require more sun exposure than other vegetables. A general rule is that the container garden needs at least six hours of sunlight per day. Many plants may require more. The fruits of the garden like tomatoes as well as eggplants, cucumbers, and peppers need sun to ensure the optimal development. Root vegetables like beets, carrots and turnips, even though they could require less sun can produce the most productive harvests if they are planted under full sunlight.

There are crops that can be grown under partial shade. The leafy varieties of vegetables, such as kinds of greens and lettuces can be produced with just 3 hours of sunshine per day.

Southern exposure will offer your plants the maximum amount of sunshine and warmth that is possible. Western exposures or Eastern exposures should be the next choice. A Northern exposure is the most efficient in terms of quantity of heat and light.

Follow the Sun

If you are deciding where to place the base of the containers, be aware at the sun's rays for several days. Note the location of any trees or buildings or other structures and also where the shadow falls at any time of daylight. It's not enough just to have a south-facing exposure. you must also be mindful of shadows that may stop your garden from receiving the maximum amount of sunlight for the most effective outcomes feasible.

Wind is a crucial factor when choosing the site for your backyard. I've witnessed many containers fall over due to the breeze. There have been plants that were shredded in the winds just during the fruiting season. If you're in the midst of an enclosed wind tunnel or strong winds that are raging through your property Be sure to be aware of this and set your containers in a secure location away from the windy zones.

Chapter 3: Make Your Containers Movable

Large containers for planting is likely to weigh in at least 100 pounds or even more. In order to make the container more mobile and easier to cleaning, platforms and dollies that

have wheels or casters may be used to help move containers from one place to another in the case of growing your vegetables in pots. This can help transfer garden containers in the event that you need to move them or if the place needs change in order to benefit from the motion of the sun, or to shield the plants from winds storms, hail, or other weather. This can be useful when it comes to autumn, in the event that you wish to move the containers to your garage or another protected area to shield them from early frost.

Make Sure Water is Accessible

Another aspect that must be considered when planning your container garden is the availability of water. Containers require plenty of water. If you are lugging around buckets of water throughout the all day, you'll get tired of the project and conclude that it's simply not worth it.

Container gardens need lots of water. The tomato plant that is fully grown within a 5-gallon bucket will require a gallon per day. It's not necessary to commit a significant amount hours weeding and tilling However, you should remember that pot gardens require watering frequently. It's not a good idea to wait until tomorrow since you'd like to head to the beach now. Be sure to drink water prior to leaving otherwise, you'll see that stunning tomato plant dying If you don't water it for all day. The containers that self-water need to be filled with water with water, so make sure you find them close to the source of water.

A hose with a long length that is able to be pulled onto the reel of a garden hose is the best choice. When you're at the hose section be sure to check out the watering sprayers. There are numerous types of nozzles available, so make certain to choose one that can provide a large selection of sprays for different purposes, including irrigation, misting, or fertilizing.

As your vegetable begins to attain their maximum growth potential You will notice that many species like tomatoes, peas and green beans require support. It is possible to place the plants near walls or structures, and be aided by wire string or Trellises.

It is also possible to provide assistance within the container. A tomato cage is a good option for pepper and tomato plants. Pyramid or cone shaped supports are more effective over flat ones. Also, you can utilize stakes made of wood. A wooden stake placed on the edge of the container in order to create an elongated tepee also can be used. When you put your support pieces inside the container when your plants remain tiny, you can be able to avoid harming delicate feeding roots in the future.

Maintaining the health of container plants with the right soil care

Making the choice to plant a container garden is among the most important and fundamental decisions that gardeners who work in container gardens have to make. Container gardeners who are new will likely take a trip outside and put in the gardening soil. Big mistake! A majority of garden dirt has clay that is dense and blocks soil from draining properly. In the event that clay becomes dry, it can't withstand water and the dirt tends to slide out from containers. It makes watering challenging and messy task.

Clay cannot support the growth of plants in a healthy way and shouldn't be considered in the selection of containers for gardening.

The other yards might include sandy soil. It requires lots of conditioning to retain sufficient moisture to support container plants. When you have added peat, compost and vermiculite as well as other ingredients to the soil you have created however, you may not have the right mix that can thrive in the container you are planting in.

When you have a garden that is heavy it is difficult to find room between the soil particles. Due to these tiny spaces in which plants grow, once they are fed, the water pulls the soil to remove air through filling the tiny space. The quantity of air that remains in the soil following water drainage is the main factor that determines the growth of the plant as its roots require air to allow to breathe, survive and development. It is evident that with garden soil, it is likely that at most, you're not getting optimal yields from your plants. The worst-case scenario is that your plants might be slowed down and die due to insufficient air.

It may seem complicated however, choosing the right the right soil for your container garden needn't be a major choice. The most important properties of soil for successful development should include:

The water is quickly drained through the soil

* There is enough air in the soil following drainage

* A reservoir of water that remains in the soil following draining.

Things to Consider when choosing a Container Gardening Soil

If you're a novice and are looking for commercial potting mixes may be the best choice. The mixes typically comprise of two elements, organic as well as mineral. Organic ingredients could be peat moss or bark from hardwoods, shavings or pine bark or fir bark.

Minerals are typically in the form of vermiculite or perlite. Sand can also be used for aeration and as a means of rooting cuttings.

Vermiculite (Terralite) is akin to mica, When crushed and treated with heat, the flakes of mineral grow to 20 times their size at the time of their initial. Perlite (sponge rock) is a rock that has volcanic origins. If heated, it pops as popcorn, and similar to vermiculite it expands 20 times the volume of its origin. The main difference between vermiculite and perlite lies in the manner minerals store water. It is a fact that water remains within the flakes of vermiculite, whereas perlite holds water within the granules. As a result, perlite will dry more quickly than vermiculite.

There's been a great deal of debate about vermiculite. It was in the 90's when the mine located situated in Libby, Montana that supplied much of the vermiculite used in the US was closed because the rocks that vermiculite was sourced included asbestos.

In the present, the majority of the vermiculite originates from mines located in South Africa and China produce vermiculite which is appropriate for domestic use. Vermiculite and perlite can be classified as organic, or not organic. Also, organic vermiculite is available made by Esponsa. If you're looking to learn more about the controversy around the usage of vermiculite for gardening, take a look at this post "Is Vermiculite Safe to Use for Vegetables?"

The basic "soilless" mixes are free of weeds, diseases, as well as insects, and are ready for use. Gardeners often purchase commercial mix after which they go home to mix in garden soil in order to "stretch it out." It is completely in contradiction to the idea of commercial mix. If you mix in soil, you eliminate all of the benefits that come with the sterilized mix. Are you truly ready to get started fighting off pests, weeds and disease within the pots you have? Take a look before you add backyard soil into your mixture.

In addition to these benefits Commercial mixes contain all the necessary nutrients to start plant growth inside the blend. Thus, these mixes are in good shape for growth immediately. Mixes like Jiffy Mix, made of vermiculite and peat moss also have the benefit that they weigh less. It could be an important benefit if you are planning in moving containers or would like to put in window boxes or hanging containers. If the garden you are planning to plant is on a roof, balcony or window box, this can make some difference to the way you design your garden.

Chapter 4: How Much Commercial Potting Soil Is Enough?

If you buy a two-cubic-foot bag of commercial potting dirt it is possible to plant these:

* Containers of 20-22 gals

* 35-40 pots, 6" deep

*If you buy the 4-cubic-foot bags, you will be able to make a planter container with 24" x 36" x 8" in depth.

The general rule is to be sure to follow these tips to determine the pot's size (insert the table in your bulb book)

Using Homemade Mixes

After gardening for some time some gardeners choose to create their own soil mixtures in the event of choosing a container gardening soil. It is as easy to find different recipes as gardeners are available to mix up your personal.

A general potable soil blend includes

* Sterilized soil. You could purchase it in bulk, or you can make your own (Directions are provided)

* Sphagnum, peat moss or sphagnum

* Vermiculite or Perlite

* Compost

* Bone meal

* Polymer Crystals (optional)

General Use Potting Mix for Container Gardens

Three parts of peat

* 1 part sterilized garden soil

* 1 Part aged compost

* 1 Part bone meal

2. 2 Parts perlite

* Crystals of polymer if you want (follow instructions on the packaging)

How can soil be sterilized?

If you're thinking of making your own potting soil There are a variety of methods to do this. Take a look at the following links to find out what method you're interested in trying as well as what your preferred method might be.

http://www.gardeningknowhow.com/garden-how-to/soil-fertilizers/sterilizing-soil.htm

http://voices.yahoo.com/four-ways-kitchen-sterilize-soil-11748054.html

The ingredient that truly does the job

As a child an experienced gardener offered me these tips to help the new seeds and plants to grow and thrive. I'm passing it onto you because I've observed this method to perform extremely well. It is a good rule of thumb to make use of the following ingredients to plant all plants and seeds:

Mix them together:

* 1 cup of sugar

3 Cups bone meal

* 1 cup Epsom salt

Directions: Pour some into the hole prior to planting seeds. My experience has shown me that it gives seeds and plants an excellent start, and promotes the growth. It's hard to explain why because I'm no scientist however I can inform you by personal experience that it does a great job.

Tips for Selecting Plants

In selecting the plants you want to plant in the container, select those with healthy leaves and a strong root system. Find plants that are small and balanced. Beware of plants that have slender branches, loose foliage or the roots that are densely packed. Get the plant out of the pot, and examine the roots. Are they bright and white or deep and shrunken? Don't plant plants with excessive growth, which aren't attractive as they develop.

If you're planning to spend the time establish and maintain the container garden, do not risk your garden to disappointment by purchasing plants that have been rejected by others and sales items that are past their best. Don't forget, it's not one if the item isn't useful or valuable to the person buying it.

Consider your garden's plan when you are buying plants. If the tomato you are in love with is a climber, and you don't have a way of assisting it in climbing to the top, it's going to be messy. If the container you have is only able to hold three plants, you're not gaining money by buying an extra quantity since prices are lower.

Check the duration of the time needed to grow in the area you live in. Even if the nursery stocks the specific variety of squash, they're not able to guarantee that your season will be sufficient for the plant to yield the fruit. They purchase from national sellers and in many cases the time frame may not be sufficient to produce the crop you want in your local area. It's a case that buyers should take care.

Trust me when I say you'll be swept off when you are looking for plants.It's easy to buy too many the plants you want to grow in your containers and especially when they're at their peak of beauty. Be sure to make a list of your plan for the garden when shopping. You may discover an improved variety or different vegetable that you'd like test, but if it's necessary to replace something else, you can cross a item out of your shopping list.

If you purchase too much it, you'll have to buy additional containers, fertilizer, soil, longer time for transplants to the next location the list, and on. The idea is obvious.

Prepare for Soil Drainage When Planting a Garden Container

After you've cleaned and disinfected your pot, your next task is drainage. The old method that involves covering pot drainage holes with rocks or pot shards in order to assist in the drainage process and prevent soil from escaping from the holes is currently being challenged in certain gardening circles.

In the event of covering pot drainage holes by shards of pot or stones is only going to allow less soil for the root system of the plant. An alternative method of preventing soil from escaping is to cover the hole with a piece of steel mesh or coffee filter to line the drainage hole. If you find that you require to boost or enhance drainage, place in a wick and allow it to stretch over several inches lower than the surface of the pot. This gives your plants the opportunity to develop more soil within as well as allowing greater air access for the root system and the roots of your plant. After drainage has been completed then it's time to add pots and mix. Then, you can put your seeds in.

Place Your Container Before Planting

The majority of vegetables can be grown in large pots. They can accommodate 40-50 tons of soil. When the plant is planted and the soil has been well-watered, the containers will weigh a lot. Consider which location you would like the container to go before planting it.

Place your container in its position prior to the planting. If you're unsure the best place to put the container, place your container on a dolly that has wheels prior to planting it.

Seeds or Plants?

Start the vegetable garden using seeds or seedlings purchased from the garden center or nursery. If you opt to plant the seeds yourself there are instructions in this e-book for free: "Timely Tips for Starting Seedlings at Home": You can download it instantly the book from Texas AM Extension to be highly beneficial.

If you opt to plant inside your container make sure you be attentive to the seeds. The carrots, greens, and radishes among others are likely to sprout in close proximity to one their counterparts. If they aren't removed early enough, they'll crush each other and become too congested for them to live.

Choose Seeds over Seedlings When you Have a Choice

In the case of growing vegetables in pots, you'll see that lots of vegetable seeds can be planted inside the container. They develop from seeds immediately without the need for transplants. This includes lettuce, radishes carrots, beets and carrots cucumbers, green beans melons (if your growing season lasts sufficient long) as well as squashes.

I've witnessed gardeners who are new paying $1.00 to buy one cucumber seedling with only two leaves. If I can get the chance for them, I advise them to take the seedlings away and purchase a bag of cucumber seeds at $1.00 and advise that they should plant the seeds inside the pot and not plant the seeds. Most of the time, seeds expand faster than seeds because they aren't subject to the fear of transplanting.

Do not plant seeds of vegetables far enough because this forces young plants to push them to reach the surface. They will be depleted of the vital energy they require to flourish and develop over the top. Follow the depth of planting listed on the back of your seed packet.

Naturally, if need to be anxious and require an attractively planted container right away then feel free to select seedlings instead of seeds. Your choice is based the budget you have and your preference.

Chapter 5: How To Pot Up A New Plant In A Garden Container

When you are planning to plant a new flower make sure it is properly prepared by placing the plant into a bucket of warm water. Keep the whole pot submerged to the water's level for around a minute until the air bubbles are dislodged in the potting mix, and the soil is completely damp. By using this method, you can ensure the soil is moist, and aid in adjusting to its new environment. Also, it will make taking the plant out of the pot much easier after you have transplanted it.

If you're employing a terra cotta clay pot, make sure to soak the pot in water for about 10 minutes prior to plant. Terra cotta's porous nature means it will ensure that the pot doesn't absorb all the precious water you provide to the soil after you transplant the plant.

If the plant's roots are sprouting out of that drainage hole it must be relocated to a bigger container. The container you choose is one size larger than the one is currently living in. If your plant is located in a 3" pot, move up to a 4" pot. If the plant is in an twelve" pot, the next size would be a 14" pot.

The new pot should be filled with soil to the fullest. The original pot should be pressed into the mix of potting soil to create the form and size of the plant as well as its roots. This gives you an exact idea of the space it will require for it to flourish at its new location.

Take the plant out of the pot in which it was originally placed. Make sure you don't remove the plants out of pots. Turn the pot upside down and then tap it on the wall. Keep the root ball in place by putting the stem of the plant in your hands. Release the soil that is tightly packed along with the root ball. After keeping the plant in place and then lower it down into the new pot. The new soil should be firm around the edges of the ball. Fill the soil and building the soil until it is at the same height as the soil surrounding the plant.

Make sure that the plants are placed within the container in a way that the plant's highest point is the same height that it was in the original container. No higher and no lower.

If the container is big you should cover it layer-by-layer with soil, then press it down gently using your fingers. This can prevent compaction of the soil which could block drainage.

The addition of Polymer Crystals in the soil mix can be an excellent idea. The crystals absorb water then let it out as the soil is dried. These crystals reduce your watering time, and will help to ensure your plants remain moist in all seasons.

It is essential to water the plant right after it has been put in. Examine the plant every 15 to 20 minutes. If there is an under-plant saucer take out any water that exists to avoid the root from rotting. It is suggested to put a mulch made of bark pieces as well as marble chips or perhaps a groundcover like alyssum or in phlox to cover the top of a big container.

It will not only make the container look nicer and reduce the rate of evaporation but also prevents the soil of your garden from getting damaged when you water the container.

Put your container into an area that is protected for at least a few days so that the plants can adjust to their new surroundings. After a couple of days after that, you can relocate the planter into the open space and into bright sunlight.

How to Water Containers

Understanding how to water your container plants is among the most important essential skills to master. A lot of people say "I just don't have a green thumb" probably aren't aware of the how of watering their the plants.

Most likely, they are overwater until the plants become so that they are flooded, forcing air into the soil and the roots of the plant have no food for air. In this situation, the plant literally is submerged. Another scenario is that it is that they completely disregard the plant until it ceases to function due to water shortage.

The majority of people aren't born with "green thumbs". The best gardeners aren't out of the ground, and nothing can replace knowledge and experience in what to do with watering container gardens.

There are numerous factors to take into consideration when beginning to learn how to water containers. Below are a few basic elements that can influence the way you take care of your plants.

Container Type

Containers made of Terra clay, terra are porous, and can be described as "breathable". Potted plants in these kinds of pots will require frequent watering than those which are housed placed in metal, glass, or other non-porous plastics.

Soil

If the plants you plant are potted with heavy, thick garden soil, it will require less frequent watering than plant pots that are made of industrial "soil-less" mixes.

Light

It is obvious that plants placed in the sun can get dry faster than containers that are in shade.

Container Size

The dimensions of the container will determine how often it is necessary to keep the plant kept hydrated.

Temperature

On sunny, hot days, you might must water your container often throughout the throughout the day. Make sure to check your container at least of every other each day. An established tomato plant in full sunlight will need more than a gallon water per day.

If you are watering your container garden it is recommended that the water be warm and room temperature. The cold water can shock plants' roots, which will slow the development of your plants. Avoid using a direct stream of water that comes from an water hose. The forceful stream can clean the potting soil out of the pot. Make use of the "water-breaker" or sprinkler nozzle or a watering container that has a spray spout gentle water your plant pots.

Sometimes, you will need to feed dry plants the water they need several times to allow the soil a chance to soak up the water. Give it a bit of water, then take it for a minute or so, and then give another amount of water. If you water your plant excessively, the water is going to run away through the bottom of the pot. The soil won't have the capacity to absorb the water in a sufficient way.

Don't let water sit on a plate or in a saucer beneath the dish. It is believed that water that sits on the lower part of the container can lead to mold and causes root decay. If you are watering your plants, make sure to return after 15 to 20 minutes, If there's the saucer that protects the plant, and then empty the water.

How to Decide If Your Container Plants Need Water

You think that you are able to maintain your garden in a specific time frame that includes Tuesday, Wednesday or Friday, you're wrong It's not! The plants must be tailored to in order succeed with your container garden. It is essential to water them in accordance with their timetable rather than yours.

The Finger Test

If you're just beginning to become a gardener who is unsure what time and when to water the container garden it is the easiest way to know if the plants require irrigation. Put your finger in the soil until the first finger (second one if container is big). If you notice any humidity, it likely doesn't require the water. Pull your finger out. If soil particles are stuck to your fingers it means the plant is doing fine and does not require irrigation. The roots require to be watered and are located on the bottom of the pot.

When you've got started your garden in containers You will discover through your experience whether a plant requires water simply by looking at the soil and rubbing the surface of the container. Make sure you water only the soil and not the foliage when watering. If you are using over-the-top watering, the water that remains on the leaves over a long time could cause mildew and mold.

What Type of Water Should I Use?

All agree that the most effective option of water to water your containers is rainwater. A lot of people make decorative containers for collecting rain water to water their gardens. If this isn't the best option for you then it is possible to use a Brita filter is a good choice to provide water suitable for this usage.

It's likely that you're asking "Why can't I use tap water?" It is possible to make use of tap water in the event that the plants you're growing can only last the course of one season. If you're cultivating vegetation such as trees, plants, or even herbs that which you intend to keep for an extended duration drinking tap water is not ideal choice for these plants. The plants are extremely sensitive to the quality of water, so the tap water you use is treated chemically, it is likely that in the future your plants could be damaged by the slow poison impact.

Chapter 6: Self-Watering Containers

Many people today try to manage too many interests and pursuits with a limited amount of time. That's why self-watering container technology will be of interest to the majority of gardeners. There are ready-made containers available or, if you're adept with the use of a few simple tools, you could create your own self-watering container.

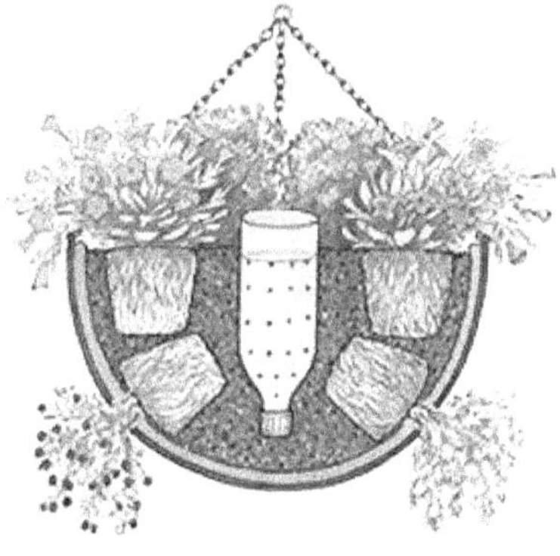

It is without doubt that the Earthbox is one of the most well-liked and widely recognized self-watering item available. It has been available since the year 1994. Find out more information about the Earthbox on this page.

If you'd like to create your self-watering container I'm able to provide instructions on this page, but because people have a variety of requirements and preferences I've included a number of websites in the Resources section. They offer a wide range of methods to make your own self-watering pots.

Watering Your Containers While on Vacation

If you've got an outdoor container garden, making sure that the plants hydrated while you are on vacation can be a challenge. An envious gardener is an ideal option, however it's not always the best option. If you do not have an outstanding option, check out this alternative list:

- Water your plants well prior to leaving on your trip.

- Move the most fragile containers and plants into the shade, then cover them with a clear plastic tarp.

* Pour small containers of food into an enormous tray of compost, sphagnum or sphagnum.

* For brief durations small pots are set on pebble trays and covered by water.

* If you are planning to stay longer put your plants in tray that are lined with Capillary Matting. This matting can be found in your neighborhood garden store or even online. .Arrange the matting such it is resting in a sink, or a bowl covered with water.

The larger pots which require a lot of effort to transport are able to be kept moist using capillary wicks that are made of mats or towels. Put part of the wick in the clay soil in the pot, and let the other end hanging into an empty bucket. Also, you can make use of a rope made of cotton as the wick.

* For costly and scarce collection of plants It is recommended to purchase an irrigation system that drips in containers

If you're skilled with a tool, look at the Resources page for directions for making your personal drip irrigation systems.

Fertilizing Tips

Fertilizing your container garden can seem like a hazard. What is the best time to give my plant a feed? What kind of food should I feed them? Are different plants able to use different foods? This is a question gardeners frequently ask themselves. If you're just beginning to become a gardener in containers, you should be certain to go through the article below for a better understanding of the way to read a label on a fertilizer.

How to Read a Fertilizer Label

If you've ever looked for fertilizer, it's likely that you've become completely confused by the vast variety of fertilizers that are available. The FDA provides standard guidelines for labeling to assist you in understanding the different fertilizer ingredients. The majority of commercial fertilizers include 3 numbers printed on their front labels that are which are separated with dashes. As an example, 5-10-5. It is the analysis of fertilizer or weight percentage of the three main elements that plants require: the phosphorus, nitrogen and potassium and potassium, in this order. They are abbreviated N-P-K.

The primary ingredient is Nitrogen. It encourages growth of the foliage, in addition to other advantages. A fertilizer that is 5-10-5 will contain 5percent nitrogen in weight. The next ingredient on the list as Phosphorous. The ingredient is essential to the rooting system of plants and the setting of buds.

The 3rd element is Potassium. The mineral is essential to the general health and strength of plant life. Other ingredients are typically included on the back of the container. There is no one type of fertilizer that works for all plants, therefore typically you'll be searching for an extra boost in ingredients.

How Often Should I Fertilize?

If you're planning to plant your containers using a commercial mix, all the necessary nutrients required for the growth of your plants and overall health are included in the mix. The mix should last about 3 to 4 weeks. Since watering is a method of removing fertilizer the more often you use water is the determining factor in how frequently you fertilize the plants of your containers.

If you are fertilizing your pot plants It is important to understand the needs of the plants you are fertilizing. Certain plants need an alkaline soil whereas others like the acidic soil. One way to find out the preference of each plant is through research and studying about the plants you've selected to work with. The information you need is usually found on the label of the plant or the seed packet. However, in the event that it's not mentioned, it is possible to search for the specific plant's name through the Internet or even in gardening guides.

If you're making your own soil mix or if your container is established for some time it is possible to conduct an soil test to assess the soil's pH. If the pH of your soil is excessively high or low, plants won't be in a position to absorb certain minerals, even if they're present within the soil. The pH Tester can be bought online or from your neighborhood garden centre.

A few container gardeners choose to feed their containers with the use of a water-based solution once every irrigation. If this is the method that you would prefer, apply only one-quarter of the mixture called to be applied in a regular basis. The plants in containers require higher nutrients than plants in gardens due to the tiny amount of soil that is contained in the pot as well as the regular timing of watering. The container needs to compensate for the small area of roots through a frequent and regular food and watering schedule.

It is not necessary for plants to consume a huge quantity of fertilizer at a moment, however by feeding your plants according to a scheduled time frame, you'll encourage an abundance of growth as well as a constant show of colour. Be aware that smaller pots as well as soil that is lighter will need to be fertilized more often as opposed to bigger pots that are filled with heavy soil as regular watering can remove nutrients from the soil.

Chapter 7: How To Fertilize Container Gardens

Do not fertilize your containers in the event that your soil is not moist. Any fertilizer that gets in contact with the dry roots could upset the plants. Make sure that the soil is wet prior to you apply fertilizer to your pot plants. If the roots are damp, your plant has the best potential to take in nutrients. Also, it is a good idea to feed your plants in the early morning. In the morning, as sunlight and heat grow, plants will begin what's known as "transpiration", the process release of water through the leaves. When this water evaporates and absorbed, it is further taken up by the roots which then transport the fertilizer into all areas that make up the plant.

Timed Release Fertilizer

A lot of people choose to fertilize containers using a timed release formula. After the plant has been kept watered, the fertilizer will be released in small quantities. This means you don't have to recall the last time you provided your plants with water, yet it meets your plants' need for an ongoing intake of nutrients. Be sure to follow the instructions on the label for any garden mix that is commercial or fertilizer. The over fertilization of your plants can be just the same for the plants as under fertilizing.

It is available in a variety of types. It is available in pellets that are placed in soil, tablets that are buried in the soil, or fertilizer spokes which are inserted inside the container.

Particular "Mixing Nozzles" for your pipe

It is possible to purchase a specific accessory nozzle that will pre-mtix fertilizer to the water in the process of watering plants to allow the direct distribution to plants. Miracle Grow, Rapid Grow and Hyponex together with various liquid fertilizers are all producing this nozzle for liquid fertilizers. It is an excellent time-saver if you own several containers because there is no need to mix fertilizer into a separate container.

Spotting Mineral Deficiencies

Ofttimes, regardless of how much we care of the plants we have, sometimes they aren't thriving. If your plant is suffering that has yellowing leaves, low growth rate as well as weak stems and inadequate root systems, it might be a sign of deficiencies in mineral content. It is possible to determine the type of mineral needed from observation or make use of a soil test kit at the local garden or nursery centre. Here is a table which you can refer to when you suspect that a deficiency in minerals could be affecting the growth of your plants.

Mineral

Deficiency

Symptoms

Remedies

Nitrogen

Older leaves, usually located at the base of the plant

They will turn yellow. Stems can also turn yellow, and turn spindly.

The growth slows

All manure

Cottonseed

Ammonia

Urea

Potassium

Older leaves get flecked or stained, edges turn

Dry and scorched.

The stems are brittle. The root systems are weak.

Produce ripens unevenly.

Potash

Wood Ash

Seaweed

Kelp

Phosphorus

The plants turn dark green usually turning to purple.

particularly the leaf's underside.

Fruits and flowers are not as good.

Phosphate

Bone Meal

Greensand

Vetch

Companion Planting

Planting companion plants for vegetables is a practice that has existed for a long time. It's been recognized over time that planting various kinds of plant species next to one another is profitable for them.

There are a variety of reasons for this can be the case:

The companion plants of vegetables are a great way to attract beneficial insects or keep pests away, thereby offering an effective method for the control of pests.

* They can provide shade, and are often able to loosen the root of the other's plants.

Aromatic herbs can help flavor development within their plants, like basil or the chives in conjunction with tomatoes.

It helps increase production yield by putting the same plants close to one the other.

However when you plant crops next to plants that don't match in terms of growth or yield, the outcome of your crop are likely to disappoint. Below is a diagram to reference and provide you with some of the most popular gardening tips for companion plants.

The most well appreciated flowers for vegetables is the simple marigold. They're able to repel all sorts of insects. They're especially good with the cabbage, tomatoes, potato as well as asparagus.

The other plants that possess the capability of preventing potatoes bugs include nasturtiums catnip, and tansy. The potato harvest will gain from the companion plants of these flowers.

* Borage is planted near tomatoes. It hinders the growth of tomato hornworms.

* Place a thyme plant on the edges of your tomato planter. It is beautiful flowing across the pot. It also gives flavor to the tomatoes.

Plant Likes Dislikes

Artichoke Parsley Garlic

Asparagus Tomatoes, Parsley, Basil None

Beans Celery, Cucumbers Onions, Fennel

Beets Bush beans, lettuce Onions Scarlet Runner (and other pole beans), Mustard

Broccoli Potatoes, Thyme, Sage, Rosemary, Dill Strawberries, Tomatoes

Cabbage Onions, Dill, Celery Tomatoes, Strawberries

Carrots Onion, Garlic, Leeks, Tomatoes, Lettuce Dill

Cucumbers Lettuce, Peas, Sunflower, Radishes Potatoes, Aromatic Herbs

Garlic Carrots, Beets, Strawberries, Tomatoes, Lettuce Peas, Beans

Lettuce Carrots, Cucumbers, Strawberries, Radishes, Garlic Fennel

Onions Lettuce, Beets, Strawberries, Tomatoes All Beans, Peas

Peas Turnips Mint, Radishes The entire Onion family

Potatoes Beans, Cabbage, Corn, Horseradish None

Radishes Most plant species, but especially Leaf lettuce Cabbage Family and turnips

Spinach Peas, Strawberries, Rhubarb None

Squash Fennel, Nasturtium, Cucumbers None

Tomatoes Asparagus, Carrots, Celery, Onion Cabbage, Cauliflower, Fennel, Dill

Vegetable Container Gardening With Kids

A lot of children live in urban environments and the only time they come across vegetables comes from the store, or possibly the farmers market. Many children are unaware the roots of carrots are rooted in the soil, as cucumbers are cultivated on a plant. As an infant, I was actually thinking that there were pickle trees!

Gardening in containers can be a fantastic method to get children involved in gardening and teach them about the process of growing things. When we were children were young, they loved having their own garden with any vegetables they could think of. Most popular was watermelons (yes I'm aware this is a fruit, however it was a staple in our garden of vegetables). The kind known as "New Hampshire Midget" is small watermelon which can be grown in container gardens.

Children being able to choose to plant and manage their own container, makes the opportunity for great lessons on accountability that are as taught by adults and absorbed by kids. Encourage them to choose quick-growing plants since, as we realize, children are looking for quick outcomes. It is also helpful if they select foods that they like eating. After children have eaten an oozing, warm tomato or clean off and eat a freshly-digged carrot, they'll be stunned by the change in taste between the supermarket items in the store. It is possible that they will be pleasantly shocked to learn that some of the foods which they thought tasted bland or tasteless actually are delicious and sweet.

The cultivation of herbs and vegetables can be a fun family project that you can enjoy with your children. Planting vegetables in upside-down containers appears to be most children's favorite method of growing. be fascinating. Imagine their excitement when they see plants grow right in front of their very eyes. And upside down! What a great and educational activity with the benefit of having a good time with each other!

Chapter 8: Controlling Pests

Pests don't pose any problems when you plant your garden in containers. If you've taken care to clean your pots thoroughly and use only clean composted potting mix that

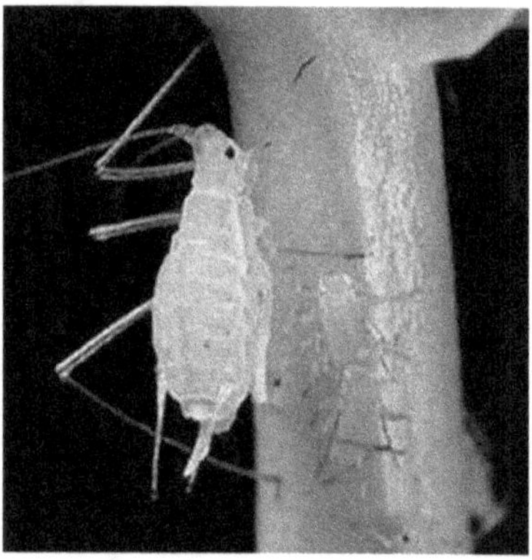

has been cleaned as well as burn or discard the foliage from last year You can spend all season without encountering issues with pests or diseases. Since we're constantly making sure to water and check the vegetable gardens and weeds, it's easy to recognize and address the issue promptly.

Below are a few the most common pests that you'll encounter along with the solutions they provide:

Aphids

Aphids, often referred to plants lice, are among the garden pests that cause most destruction. They feed on sap that is found in many plants. They're soft with bodies that are available in a variety of colors, such as brown, black or even pink.

If you see only just a handful of insects, use q-tips that have been coated with rubbing alcohol to get them off the plants. While aphids can be a major issue, they are able to be eradicated by spraying plants using the hose whenever they first appear. If that doesn't work you can spray them with this recipe for homemade aphid spray:

1 quart of hot water

* 1/2 tsp. liquid dish detergent

* 1/4 tsp. Cooking oil

Combine this using a spray bottle. spray the plant every day until aphids go away. It is also beneficial to treat whitefiles. If you find that the problem is serious, you might want use the use of commercial insecticides, however when you're growing veggies you should avoid this method.

Mealybugs

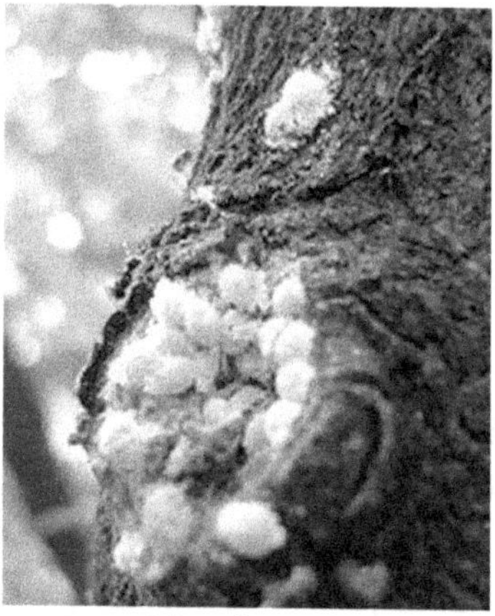

Mealy bugs resemble the white color of mold or cotton on your plants. Mealybugs, just like aphids consume the sap of plants. They are able to be controlled through using an insecticide specially designed for mealybugs.

Slugs, Snails, and Caterpillars:

Damage can be identified from caterpillars via the holes that they make in the leaves. The caterpillars can be large enough to remove manually. Most likely, you will not encounter problems with hanging baskets but sometimes you'll see them adore the window of your container. If this happens then apply Vaseline on the edges of the container. This keeps them out of the containers.

If you notice insects that you do not recognize or have a plant that is unhealthy that is yellowed, damaged or swollen leaves and you do not know what's wrong is, you can take a photograph on your smartphone, place the damaged plant or the insect if want, in an airtight bag, and then go to your local garden or nursery. The people who work at these locations are experts in their field and are available to assist you in bringing your plants back to their original health and vitality.

Like I said previously, the best way to guard your plants from insect damage and diseases is to keep them clean. Make sure that your pots, soil and plants tidy and clean and they'll surely provide you with plenty of harvests with very little trouble.

Saving Seed

Through the years seeds have increased in cost, but have decreased in quantities of seeds bought. Because of this as well as because many gardeners have concerns about GMO (genetically altered organisms) A lot of gardeners are experimenting in saving their own vegetable seeds.

It's an excellent idea, however it must be from heirloom seeds, not hybrids. "What's the difference?" you may ask.

The Heirloom Seeds are those which have been cultivated over generations, and handed down from gardener to gardener over long periods of time. They are pollinated naturally by birds, wind and bees. They may not be so appealing or possess more yield than hybrids, nevertheless, many gardeners enjoy their taste to the natural flavor of seeds that have been the favorite by gardeners for a long time. Also, they enjoy the satisfaction of preserving our garden heritage in the form of preserving foods enjoyed by our early ancestors.

Hybrid seeds can be cross pollinated from one the plant to the plant by farmers in order to encourage larger sizes larger yield, higher quality, and color. Hybrid seeds are extremely stable and are able to be relied on to grow crops precisely like the ones advertised in seed catalogues. One disadvantage of hybrid seeds is they are unable to be stored and have to be purchased each year in order for the exact same result.

If saving seeds and eating delicious heirloom veggies appeals to you, then you'll get all the info that you require on Seed Savers. Seed savers exchange is non-profit group that is dedicated to preserving and sharing of the seeds that are heirloom. Membership costs $30.00 annually, however it's worthwhile if you wish to keep seeds in your garden and swap them among other members. Seed exchange is absolutely no cost. The seeds are available for buy from their store online without having to join.

Each vegetable has its own demands when it comes to saving seeds, I'd need to create an entire guideline to cover justice to this topic. However, check out this page Seed-Saving-Instructions for comprehensive instructions on the proper way to save and store seeds for your next gardening season.

End of the Season Clean Up

• At the close each season dispose everything in the pot. Make sure to not add this in your compost bin and do not reuse the mix of potting.

The vegetables are an important source of pests and illnesses. Do not take the chance of bringing on any diseases or pests in the following growing season.

Make sure to disinfect the containers you use. The sterilized containers can help you cultivate healthy, strong plants, with no issues with pests and diseases.

Sterilizing Your Containers

An excellent habit to fall into is to clean your containers towards the end in the season of growth. It is much easier to clean your containers when you're in the full cleaning mode of an autumn day that is beautiful. Don't be an irritant when it comes to this task.

I'm aware of where I'm coming from. I've sterilized several pots in the winter's gloom and created a huge mess in my garage. I've sterilized my pots outdoors during frigid, cold March days and then frozen my hands while sterilizing dishes in the kitchen, and increased the time to clean even though I could have completed this task outside during a beautiful, sunny autumn day.So don't put off doing it - just get it done!

Take out your container of soil. Look over fragile terracotta or ceramic containers for any cracks. If you notice cracks, and you love the ceramic pot, you can sterilize it and repair the fracture using a silicone sealant. If you're making use of the pot to cook vegetables make sure you check the labels for ingredients that are toxic. There are secure silicone sealants from the pet and aquarium shops. If the container is cracked too severely for repair, dispose of it.

Use gloves to scrub the pot using an abrasive cleaning brush. clean it using 10 percent chlorine bleach solution, both on the outside and inside. Clean the pot with clear water for two times in order to eliminate any bleach remnants. Dry the pots completely. The moisture can encourage an increase in bacteria, especially if the pots are placed in stacks for storage prior to being totally dry.

Chapter 9: The Plot Of Vegetables

The size of a vegetable garden size be? This is an issue that's frequently asked by novice gardeners. There's no correct or wrong method to go in determining the dimensions of your vegetable garden. It is suggested to start with a small garden and then expand the area when you progress.

The ideal size of a vegetable garden for beginners is approximately 10x16 feet. This size of plot is enough provide a meal for the family of four in the duration of one season (through the end of summer) and you may receive a little extra, which could be frozen or given to friends and family.

If you are digging your plot, create it into an oblong or a the square. This will allow you to access the middle of your vegetable plot. You won't be standing on the plot, and then putting the soil.

The size of your vegetable plot also depends on the amount of space that you are given.

Once you're ready for sowing - you'll be able to create rows in which you will sow seeds. Try to make the rows of vegetables from the north towards the south in order so that you can benefit from the sun's. However, it's not a problem that you don't operate our vegetable gardens between east and west in order to have a great crop of veggies.

TIP: You may also plant or grow within squares. This can be helpful to use

(available at garden centers). Some vegetables include: (available from garden centres). Test it out and decide what you like to cultivate.

TIP: If you intend to get started with cultivating tomatoes and cucumbers, it is possible to grow them in containers, pots or even in grow bags.

Important: Vegetables do more effectively in warmer climates and need lots of sunlight to produce productive yields. Therefore, if you are able to select a location that receives sunlight for a long time every all day.

Beds raised

Raised beds are ideal for smaller plots as well as large ones alike, to plant vegetables.

Beds with raised beds are getting very sought-after due to:

* You are able to make them to the size that is suitable for your backyard.

It's easier to put in the back.

• Easier to keep clean of weeds.

It is possible to purchase them already designed and build them at your home.

They come in various dimensions and shapes, packaged as well as based on the kind and type of beds you purchase the process can range between 10 and one hour to unpack the bed before creating an elevated bed.

It is possible to find a shape and dimension that will suit the garden you have.

My backyard garden, 2011 the making of raised beds

We've made our own raised vegetable beds. These raised beds for vegetables are very simple to construct and are a fantastic option to plant the vegetables you grow yourself. It is possible to put an elevated bed into every corner of your garden.

There are many kinds of planks to create raised beds made of wood. There is a salvage yard to pick up an old piece of timber to use to build a raised wooden bed.

You can also purchase treated wood. It is also possible to purchase pre-made elevated beds made of wood. What you must do is to hammer these posts into the ground. Then you'll be prepared to cover the raised bed with dirt.

We built our own wood raised beds out of 'dung boards These are treated long boards used by farmers for holding "dung". The boards measure 15 feet long. The raised beds we built are twelve feet long and three foot wide. This means we require 2 boards for each raised bed. Each board costs PS15.00 which is about $25.

How to construct an elevated bed is:

* The planks you want to use

3" Galvanised nails

8. 2"x2" with 12" stakes

* Mix of soil (use an assortment of topsoil compost, topsoil

soil).

(Below below - our raised beds right when we put them in.)

Put the planks into the space you intend to build the beds. Set one end bit against the length of the plank, and then hammer into two nails. The same process is done on the opposite side. Then, go to the opposite side, and repeat the process at the other way. Now you have an oblong or a square bed.

Set the raised bed in its place. Then, hammer the stakes in the ground at every side of the bed. Grab a couple of nails, and smash them through stakes, starting through the interior that of the bed, into the wood planks. Make this process for every corner.

The size will depend on the raised beds, put a few stakes down each plank in order to help the planks stay in place. Then, add additional nails. (This can stop the side from bulging). Then you are able to fill the bed raised with soil.

There are scaffolding planks that people can use rail sleepers, or simply wooden planks from their collection. It is possible to use any material however, you must be aware that any treatment that is applied to the wood may affect the soil that you put in the elevated bed.

When you've made the raised beds with wood, you can add soil. The best soil to use is top soil, however if top soil isn't accessible, you may use the soil that you own. After that, you'll be able to add manure or compost (or both) to enhance the soil's structure. In addition, you can also include grass clippings, which can also help improve the soil's structure.

Raised beds help prevent compaction of soil They also offer good drainage, and also act as an effective barrier against snails and slugs.

A raised bed into an area within your yard. This will allow you to grow your own vegetables without having to dig into your gardening area. The raised vegetable beds are simple to build and a wonderful option to plant your own veggies

TIP: You may build an elevated bed. However, take care that the treatments that is applied to the wood may cause soil contamination when you add to your raised bed.

There are people who have utilized blocks, bricks or stones for a raised bed.

Chapter 10: The Soil Types Are Six

It is an vital to the growth of vegetables. There are six major kinds of soil: Clay Sandy, Silty, Peaty, Chalky, Loamy. Three of them: Clay, Sandy and Silty are typically encountered in gardens.

How do you determine which kind of soil you're in. These are some tips that can help.

What makes clay soil unique is:

* Drains poorly

Feels sticky and lumpy particularly when wet.

* Rock-hard after drying.

* Extremely slow to warm up during spring.

A hefty soil is required for cultivation

However, if drainage has been enhanced for clay soils that holds the nutrients well. You can then grow all kinds of vegetables.

Specific characteristics of Sandy soil include:

* Free-draining soil

* It warms quickly during the spring

* Dries out quickly

• Very easy to care for.

The soil in sandy areas can have the appearance of a deficiency in nutrients since these nutrients can easily be eliminated through the soil particularly during humid and rainy.

Specific characteristics of soils that are Silty include:

* It feels smooth and soapy

The soil is very well-drained - which holds in water

* High in nutrients and much more fertile than sandy soil.

It's a fantastic soil choice because it's easy and pleasant to work.

Specific characteristics of peaty soils are:

* It is a rich source of organic matter

Very only a few substances

* It warms very fast during the springtime

Peaty soil can be a great source of development of plants if it is fertilized.

Particularities of Chalky soils are

This soil is highly alkaline and has an alkaline pH of 7.5 or higher.

* Free draining

* It can be stone.

The soil of this type can have a shortage of certain minerals. This could lead to low vegetable growth as well as green leaves that turn yellow. This can be rectified by adding fertilizers to the soil.

Specific characteristics of Loamy soil include:

The * is best soil

* Good soil structure

It is a good draining soil, but also retains moisture.

The soil of Loamy is packed with minerals

• Very easy to care for.

It warms rapidly during the spring and due to due to the soil's great structure, it does not dry out during the summer months. This is the ideal soil you can have.

Testing your soil

Before you are able to test your soil, you must determine what kind of soil you have. If your soil is dry, it will require watering only a tiny space. The surface water disappears quickly in the case of sandy soils, however the water will last for longer in clay.

Pick up a few handfuls of soil, and then gently squeeze it. Look at it

* If the soil seems slippery and sticky and when you take it off from the soil, it remains in its shape, it is a clay soil.

* The chalky or sandy soil can feel rough, and the soil lump inside your palm won't grow it will break apart.

The soil that is peaty can feel extremely spongy on your fingers.

The silty and loam soil is smooth and also maintain its shape after being squeezed into the shape of a ball. It'll stay longer when compressed than sandy soils, but it will not keep its shape as well as clay soil.

Once you've completed the test above, you will should take a few ounces of soil and then place it into the jar of glass. It should be filled up with water and mix it thoroughly. Allow the mixture to sit for about 2 hours. In the meantime, you can take a look:

* If you're looking for chalky or sandy soils, many of the particles sink, and that will create

a layer at the bottom. You'll notice that the water appears quite clean.

* When you use silty or clay soil, you'll notice that the water becomes less clear and you'll only see a tiny layer of soil on the top of the container. This is because clay particles require a long time to settle.

* Peaty soil: you'll see a lot of small particles floating around the water, and it appears cloudy. There's a tiny amount of sediment at the bottom of your jar.

* Chalky soil - With chalky soil, you'll get an accumulation of white coarse particles that are found on the inside of the jar, and the water will turn grayish in colour.

*Loamy soil. You'll be able to see fairly clear water, with a little sediment at the bottom of your jar Also, you'll have tiny particles floating over the top.

In order to grow healthy vegetables, it is essential to have a good soil. Every soil is pH, and the ideal soil will have the pH around 6.5

meaning that the soil is full of nutrients. Above that level is acid, and above it, the soil is alkaline both can lead to issues within the gardening.

For a pH test in your backyard, purchase an test at your local garden center or an online store.

It's not a reason to fret even if you're not blessed with the best soil today as you have the ability to enhance the soil's structure and that's an vital aspect of organic gardening. This is due to constantly adding nutrients to your soil, which will improve the soil's structure.

This is an ever-changing process is something that as a vegetable gardener organically will never give up on. This is like constantly providing nutrients to the soil, so that your soil is taking care of your vegetables you plant.

How do you improve soil's structure?

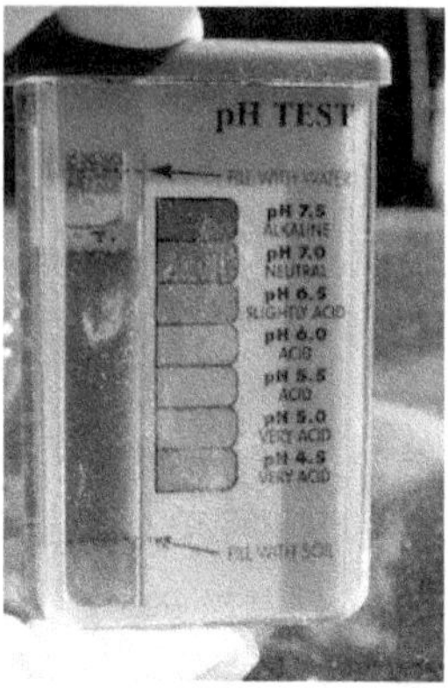

Improve your soil by digging it and including well-rotted organic matter, for example:

manure, horses and chicken manure is extremely popular among gardeners for its nutrients to soil.

It is also possible to incorporate compost to greatly enhance the soil of your garden.

If your garden is containing clay soil, it signifies that the soil is compact and tightly compacted.

Clay soil stores water, and once it has dried out, it is likely to split. The main issue in clay soil is that the roots of vegetables plant will face a lot of difficulty reaching the soil for all the essential nutrients.

This is why to make the soil work and break it down. To do this, you must add sharp sand, horticultural grit and composted organic manure that is well rotted. This helps reduce the volume of soil, so your vegetable plants are able to expand more readily.

Chapter 11: Natural Organic Fertilizer

If you want, apply a general purpose fertilizer however it doesn't alter the foundation of the soil. Organic gardeners need to take care of your soil. You do this by incorporating manure, compost grass clippings, and organic matter like leaves.

It is possible to purchase horse manure at your local stables. Be sure to ensure that it's fermented manure, and must be at least six months old, because if it's not, it could ruin the vegetables you'll plant. The manure that has been rotted will fall in pieces, unlike fresh manure which is full of straw which hasn't yet begun to rot.

Additionally, composting is a great method to

enhance your soil since it's abundant in humus

Humus can be very beneficial for the gardening

soil can solve many of the

potential garden problems. Examples include

sandy soil is composed of large sized particles.

This implies that the nutrients are derived originate from

soil could be lost rapidly, but it is not the case soil is lost quickly

You dig up lots of humus, this is a good idea.

can help to stop nutrient from being washed

made from soil.

Humus can also stop your soil from drying This is why it's crucial to make sure that you add compost to your gardening space.

In addition, we can include grass clippings, which enhances the structure of our soil.

Clippings of grass are precisely that. The lawn is mowed and then use them

The vegetable garden is littered with clippings. We ask our neighbors for grass clippings they might have. We then use them on the plant, or we can cut them into the soil.

soil prior to having put anything in the ground.

Important: Don't use grass clippings left over from your lawn, if you've been treating your lawn recently using a weed killer or another kind of pesticide which could cause death to your vegetable plants.

Watering your garden...

It is important to consider the best way to water your garden.

It's essential to install the water pipes with a ring in order to capture the most rainwater you can.

There's a wide variety of waterbutts that are available and there's one to suit all gardeners, regardless of whether you're working with a limited area. The options include:

* Standard Butts

* Decorative Butts * Space Saving Butts * Large Capacity Butts * Large Water Tanks

The Standard Barrel Butt -This is just the

Traditional barrel-shaped water butt. They

are sealed and are offered in various

Sizes: 120 litres and the 210-litre capacity.

Decorated Water Butt There are attractive

butt that is readily available, you can also get water butts

decorative wood grain motif, woodgrain effect,

the oakwood effect, and those with a shape similar to

A beehive. They can contain anything between 110 and

220 L of water.

Space Saving Water Butt - This is an excellent butt to use if

There's not a lot of room. Certain of these

butts have been tapered, so they can fit into smaller

Space is limited and narrow. Also, you can get space

Water saving wall mounts. They are the kind that can be mounted on walls.

Of water, butts can hold any number between 100 and 250

Liters of water.

Large Capacity water Butt The types of waterbutts will hold between 220 and 250 Liters of water.

Large water Butt Tanks - These tanks offer a higher than usual capacity for storage. They also have high-end materials and appear nice These tanks are available in various sizes ranging between 500 and 1500 to 1500 litres.

Then, you can connect a hosepipe to the water-butt tape to the vegetable garden in this manner.

Another alternative is to transport buckets across in the garden so that you can give the plant water when needed, however this will require work and is tedious.

Most homes also come with a garden faucet outside and you are able to install a hosepipe and then use it to provide water to your gardening. It can be a costly method if you're connected to mains water and need to pay for each millilitre of water that passes through your meters. It's a good option to taking into consideration if your garden truly requires to be watered.

A different method of watering your vegetable garden is to reuse household water. This is water that you've been using, for instance the bathwater. Many people aren't keen to make use of it because they fear that the water might be infected by soap or shampoo.

However, we've been using our bathing water for three years, and it's been a blessing to our vegetables. The vegetables do not taste bad too. Also, you can use the water to water your fruit trees as well as the lawn.

The diverter switch in our bathroom's downpipe, which fills the water-bubbles in our home, in the event that we do not need bathroom water, we turn off the diverter switch to the system for waste water.

Preparing for the vegetable patch

The seed bed is crucial when planning to plant seeds. It is true that you could reduce your expenditure by a significant amount of cash by sowing seeds instead than purchasing plants when you purchase plants. The cost of buying a seed packet will cost less than buying plant material to plant in your gardening.

The better your seed bed, the more effective. The seeds you'll be planting are tiny. Seeds need to sprout and then pop out of the soil and grow roots, and absorb nutrients from the soil to ensure they can grow. Seeds can do this more easily if soil is quite thin.

Therefore, it's essential to create a perfect garden bed. So dig over your garden. If the soil you're working with is very dense i.e. clay, you'll have to rotate it in order to break it down. If the soil you're working on is clay-like, it is more dry and easy to work using a rake. You'll notice that any soil clumps break up. When you are working in your garden, you'll observe that your soil is crumbling.

Be sure to take care of this vital step prior to planting. The tilth of your garden is crucial because seeds must have the best opportunity to sprout. If you give your soil the proper tilt, you'll definitely give them a greater chance of success than if don't have a good tilth.

Sowing & Planting

The time to plant your garden is dependent on where that you reside in and the climate in your area, we are mostly in warm countries. spring is the season for sowing and planting.

Use common sense while plant. If you plan to plant sweetcorn outside but it is forecast to frost, not plant it because it will not survive.

An excellent way to know whether the soil conditions are suitable to plant seeds and sow is by testing the temperature of your soil. It is done using the soil thermometer. It is possible to purchase one for a reasonable price from your local garden store or through an online source. The first step is to check the temperature of your soil regularly. If it is at 4.00oC between 4.00oC and 4.5oC it is time to start sowing seeds and planting your vegetables in the soil.

There is a chance that you won't be capable of growing every plant you'd like to depending on the area you live in. If you are buying plants from a local garden center or nursery it is possible to ask the staff for advice, and usually they'll only carry varieties that will be grown in the area you live in.

Seeds for growing vegetables

It is possible to grow the vast majority of your vegetables using seeds. There are specific directions for each packet of seeds on the ideal method to plant these seeds.

The seed packets will tell you an appropriate depth of where you can plant your seeds, as well as when is the most suitable time to plant them.

In order to sow seeds, you will need to

Draw a line through the soil.

However, first you have to draw the straight line.

The most effective way to draw a straight line is create the lines in a landscape

Line: Take two sticks, and then tie string to them. The string must be long enough along the entire length of the garden you plan to plant.

1. Push one stick into the ground, in the area you intend to plant with the other on the opposite side to create an even line.

Prior to sowing, you must to firmly firm the soil and disintegrate any lumps. You can perform this by jogging along the line. Make sure to keep your feet close and then shuffle across the line instead of stamping your feet. You can also employ a rake, using the rake to hammer it up and down until your soil is hard.

Tips: The ideal conditions to walk on the ground is following a light drizzle. The clumps and lumps will fall apart much more easily. When it's really hot, it's almost impossible to walk and mow a bed until the point of a perfect tilt. In addition, if the gardening area is extremely moist it can be hard to create a clean tilth till it is dry.

In the case of sowning drills into the garden for vegetables, a garden line that can be drawn taught is the most efficient method to draw out straight drills. Drills that are shorter than 2 metres may be drawn using an axe or straight wood plank.

Follow the line of your garden using the trowel and create an area of about 1/2 inch deep You can then start to sow seeds. Sow the seeds as thinly as you can, should you plant they are too dense, cut them down after they have sprouted. After that, cover them with soil. Do not put the soil over the top, but lightly cover the seeds so you don't have to see them no longer.

Be aware that generally, bigger seeds are stronger and more durable than seeds that are smaller, which implies that they're better suited to grow in the garden.

The seeds all like warmth. However, smaller seeds will develop better when the soil is cooler or you are able to grow them in an indoor greenhouse tunnel, propagator or tunnel - which is a way to gain a 'head-start'. When the seeds are sprouted and have grown to one" to 2" tall it is possible to transplant them to the outside.

When you are ready to plant them outdoors, you'll need to dry their roots. This means you place the seeds out in a warm, sunny day for about a couple of hours before bringing them inside. This lets the seeds get comfortable with the outdoors. This process is repeated for a few days in order to allow your seedlings become accustomed to temperature and environment outside. Then, only after that, do they go outside to plant.

The reason for doing this is to make sure that the seedlings in place for planting outdoors. If you just plant them outside they will be shocked and cause them to die.

It is likely that you need to freeze the seedlings you purchase seeds from a garden centre or a nursery. They've been grown inside large greenhouses, in order to ensure they get the greatest chance of success, it is necessary to make sure they are hardened prior to planting them in your gardens.

The planting of young plants, seedlings or seeds

If you've started growing vegetable seeds in your home, you'll in the future require the planting of your naive plant.

Prior to planting your younger seeds, you must be aware of the distance that each plant needs to be placed. The appropriate distance between each plant by reading the seeds packet. Then, you can make the correct size hole to your plant.

It is also recommended to grow the plant in with the same amount of soil it is being grown.

It's recommended to drill every hole in the same way It is especially beneficial when you're growing small seeds or plants on trays or the dirt. The reason is that the roots are only exposed for a short period of time. This prevents the root systems from drying out.

If you notice any stone or other rocks within the holes you've created take them out, to ensure that your young plant is able to continue growing.

Prior to planting every young plant, you may add compost at the bottom of every hole, to allow the plant to have the best start possible. If you choose to do this, be sure to combine the compost with the soil before installing the young plants.

Make sure to make sure to water the hole lightly prior to planting the new flower. The water that is drained will be much less likely to evaporate.

Now is the time to pull the plant or seedling off of the pot or tray and prepare for the planting.

Carefully remove the plant out of the container. If there is a single new plant growing inside a pot, slowly turn the pot around while keeping the plant in place and tapping the inside of the pot using your hands. The roots will be loosened and your plant will emerge easily.

If the plant is inside a tray, then grab small pieces of trowel, and slowly move them away from the roots. After that, gently separate them and do it very close to the root as in the event of damage to the plant.

Beware: Do not try to pull plants away from the stem because this could cause damage to the stem as well as the roots.

Don't disturb the roots more than what is needed. The plant should be placed in the hole. It is important to ensure that the plant's roots are in contact with the soil in the middle of the hole. Make sure that the surface of the plant's foundation is at the same level as the soil.

It shouldn't be any below the soil's surface since this may cause an indefinite puddle surrounding the plant.

When you're satisfied with your plant, will then have to fill in the holes with soil to help it grow, rubbing the soil gently with the palm of your hand to ensure it is packed tightly and not overly tight.

Give it a lightly watering with a watering container.

The plant will require to irrigate the plant every day up to a week particularly in dry weather, following which the seedling or young plant must get planted.

Also, be aware of the temperature, as should the weather become too cold, you might need to cover your cuttings by using an ice cloche. I also use an older water tank which is having its bottom shaved off, or a an empty milk bottle made of plastic that has the bottom cut. This can aid your plant's young to grow stronger.

Important: Before you plant seeds or plants in your garden, it's a great suggestion to look up the weather forecast for next week. This helps ensure that the weather is favorable to your plants, and it will help you avoid transplanting in the event that it is predicted to be warm or cold.

If you can, plant your new plants an overcast day or day that's not overly warm, to ensure that the plants' young ones can get settled in and heal more quickly as opposed to the scorching heat of the day.

Chapter 12: The Process Of Rotating Your Crops

For gardeners, you have be aware of the rotation of your crops. While some gardeners may think Crop Rotation isn't essential - I do believe it can be extremely beneficial for your garden, particularly in the case of organic gardener.

The rotation of crops can prevent the accumulation of pests and illnesses, and permits the vegetable to continue developing.

Another reason that gardeners employ crop rotation is because certain plant species get greater nutrients from the soil than other types of plants, for example brassicas as well as lettuce. The types of plants mentioned above will typically absorb a significant amount of nitrogen from soil and can thus strip the soil of this crucial nutritional element.

Other plants like roots plants like beetroot, carrots or parsnips have less need for nitrogen and are an ideal crop to plant following brassicas or lettuce plants.

Peas and beans can also be an excellent crop to cultivate because they can retain and fix nitrogen within their root systems. They also keep nitrogen in the soil for the other crop varieties.

Some crops don't get included in the rotation of crops The crops that are not included in crop rotation are utilized in semi-permanent locations. This type of crop comprise runners beans as well as Jerusalem artichokes, which are typically utilized as screens. Asparagus, another plant which is cultivated at a fixed location since they are expected to remain in the zone for a period of about 10 years.

If you are unable to do an annual rotation of your gardens in order to stop diseases from forming, you should take precautions by adding a fresh, clean soil or compost in the vegetable garden.

Through crop rotation you let the soil heal and also reduce the risk of pests and disease getting into your garden vegetable plants.

There are people who grow potatoes in the same place repeatedly which can lead to an eelworm cyst taking hold over the patch of potatoes. If you have a cystemic eelworm within your potato patch, it will be impossible to cultivate potatoes for seven years until the pest is totally eliminated.

If crop rotation was been utilized, there would be two years between plantings.

Club root is yet another disease that is affecting the family of brassicas mainly because crop rotation isn't being employed. Also, if you utilize crop rotation, it is possible to keep these diseases at bay that could affect your garden crop.

A lot of gardeners opt for a 3- or 4-year rotation in order to make sure that the plants have ample space between them prior to when they have to move for the next one. It is possible to simplify your rotation by planting the same crops in a row, like beans, peas, brassicas and greens as well as a variety of varieties of root crops.

A simple method to perform crop rotation is to use many raised beds. If you own a variety of raised beds, then you'll be able to quickly determine which crops have been growing in the particular the raised bed.

To continue to produce lots of different crops you could require a rotation.

Garden Fertilizers

For a good crop to grow, you must fertilize your crop, no matter if you're an organic vegetable gardener or not, there are a variety of fertilizers you can choose from.

There are many options for providing fertilizer and food for the plants of your vegetables.

* Bone meal, which is the coarsely crushed bones that is used as organic fertilizer. It functions as a slow-release fertilizer, and it is

highly beneficial to use in your gardening. It is available for purchase Bone meals from gardening centers and also from other gardening stores.

* Blood, Fish and Bone is an excellent soil enhancer. It can also be used to fertilize however it lasts for an 8-month period of growing. It is also used as a fertilizer for fish, Blood and Bone is a general organic fertilizer which provides nutrients to the soil. It's an analysis of 6-6-6 meaning it feeds the soil equally in the form of Nitrogen (N), Phospherous (P) and Potassium (K) You can see it in the bags with NPK.

* Potash - also naturally fertiliser and can be utilized to boost roots, and increases your fruit, vegetable and flowering production.

* 6X Pelleted * 6X pelleted Manure It's found in supermarkets or in your local garden centre. Organic fertilizer that is 6 times more nutritious than manure from the farmyard. The manure of chickens typically has an NPK proportion of 4.3

3.2 - 3.2 between 3.2 However, it also enriches soil with magnesium as well as various trace elements. This can be utilized in your garden as well.

flowers and plants. This is a cost effective option to fertilize your landscape.

* Calcified Seaweed can be very beneficial for soil. Seaweeds with calcium contain all trace elements that are naturally present to ensure your soil remains well-nourished. Seaweed that is calcified improves soil's structure, and gives your soil the proper PH that is essential for the growth of healthy vegetables.

* Growmore is an all-purpose fertilizer with an NPK ratio of 7-7-7 - 7, which means that it is the same amount of nitrogen (N) Phospherous (P) as well as Potassium (K) it's an excellent general fertilizer, however it isn't a soil-feeder as much as other fertilizers.

The compost heap in your garden

The basic concept behind composting is to use organic matter that rots down into humus that you then can use for your garden as an organic, natural fertilizer that feeds your vegetable plants.

Thus, you could include almost everything that can decay into your compost pile. In order to make compost that is good, it is essential to encourage the materials that compost be broken down as fast as they can.

There are a few things that aid and others could hinder the material's decomposition in the compost pile. If the compost pile isn't subject to rainfall the addition of water could assist in encouraging composting to degrade faster. But it should not get too excessively wet because otherwise, the compost may turn sour. It also requires an adequate supply of air.

This is why you must regularly fork the compost, then flip the compost so that it's exposed to air. At this point, you will be able to determine if your compost is actually dry. If it is, you should add water to it right immediately. If it's moist, then you don't need to pour in any additional water.

Another way to help your compost to break down faster is to put in layers of straw and wood in between other components. These are both rich in carbon, and aid in the process of composting.

Nitrogen is a different component of the composting process. This is a result of grass clippings as well as animal manure (horse sheep, cows and chickens). Do not add too many grass clippings into your compost pile as they will quickly transform into a slippery, watery pile of slush - that will end up being not beneficial for the gardening. It's essential to maintain a the right balance of substances in your compost container.

Most important for a the best compost is having various materials available within your compost pile.

As an example, grass leaves can degrade quickly, which is then triggered by the process of decay for others in the compost pile, other materials such as twigs, or the stems that are woody of the cabbages, decay less quickly and that's why it's more beneficial to cut them into shreds or chop them in smaller chunks prior to adding the compost pile.

It is possible to add any type of material into the compost pile as you want, as long as it rots. If you have a mix of different items, it will create a great compost which will be fertile for the plants of your vegetables.

The ingredients you choose to use aren't only limited to the materials you collect from your backyard You can also make use of the cooking peelings like potato, any vegetable peelings and tea leaves, used tea leaves and tea bags (better to pull the tea bag from its sleeve before adding the tea bag) eggshells, other food items that aren't cooked. Additionally, you can incorporate seaweed, wood ash and bedding, such as straw, hay or wood-chips that are slow to decompose.

There are some things you should not include, for instance: the cat and dog's droppings, do not add coal ash It's also best to not include cooked food items and meat products since they could encourage rats to eat your compost.

It could take anywhere from few months to complete the compost

It all depends on what you put in and how fast it's going to fall apart.

If you are in need of humus immediately to improve your garden it is recommended to use fast decaying materials. However, be aware that if you utilize a large amount of grass clippings that you'll get a sour stinky messy mess. Cuttings of grass will always require additional materials to be added including straw or hay.

Chapter 13: Build Your Own Compost Bin

Anyone can construct an organic composter and it's really easy to construct.

Pallets from old buildings can be used to build one or, if you've got leftover planks, you could build one up quite easily.

Materials

* 7 lengths of two 6-inch planks each plank is cut to three feet. The planks can be cut from the wood yard, or make them into pieces by yourself. It is possible to use rough non-planed wood. If you use certain preservatives, the wood will remain longer

Four posts, each about 4 feet in length and about 2" x 2".

Galvanized nails are three" long. There will be approximately 28 nails.

Assembly

2. Take two posts and put them in the ground - then hammer the planks in the posts by using nails. There will be 2 nails on each plank. Tips: Set the nails to ensure they won't be overlapping in the posts.

You should leave a little space between your boards (as illustrated below) it will help improve the airflow in the compost pile. Pre-drilling the nail holes, which will facilitate nailing and will also stop the wood from breaking.

* Nailing the planks to the three sides.

Place the bin in its place, then push the corner posts in the ground using a sledge or hammer.

Your compost bin is set for use.

This is actually extremely useful If you have enough room to

Have a double compost bin either placed side-by-side. If one pile of compost is large, you are able to "turn it" by shoveling it into the

The second bin. You can then make use of the bin that is empty for a fresh compost pile while you continue to use another pile is being used.

Composting is completed.

Fertilizers continued

Composting your own isn't the only way to make organic fertilizer. Many organic gardeners make use of a seaweed-based fertilizer since it's a highly powerful fertilizer which breaks into pieces very fast.

Also, we gather seaweed on our beaches nearby and place it in the garden as it's an amazing fertiliser. If you choose to go this route, make sure to take the seaweed in the winter season so that it can begin to break down prior to planting your garden.

Important: When collecting seaweed at your local beach, it is important to know whether there's any kind of pollution that may have contaminated the seaweed.

There are some who worry over the amount of salt present in seaweed. If this concerns you, and it is an issue for you, clean it with rainwater and then add the seaweed to your backyard.

Natural fertilizers are also available including pine needles, peat moss, and each of which is acidic, while limestone is alkaline. Depending on your soil's pH it's best to choose one over one of the.

Manure can also be a very beneficial organic fertilizer. However, you must make sure the manure you are using isn't fresh as the new manure is extremely acidic and is not suitable for your landscape. Manure that is old can be identified when it is extremely crumbly, and it doesn't scent very well.

The process of making liquid fertilizer

It's extremely simple to create liquid fertilizer, and it simply takes just a few hours to prepare your own.

Liquid fertilizer is an excellent method to improve the health of the growth of your plants since it's an excellent food source for your plants - and creating it yourself are sure to save money too.

There's a myriad of liquid fertilizers which you could make. Here are a few of my favorite ones:

* Comfrey: Comfrey is very rich in potassium, which makes it a perfect plant for making the liquid fertilizer. It is a great fertilizer to use on tomato-producing plants in order to help it develop more quickly and effectively. The only thing you have to do is place the leaves of comfrey into water until they begin breaking down, which happens quite quickly. You can then use the liquid to spray onto your fruit or vegetable plants in order to stimulate it to expand.

* Nettle: This fertiliser is very rich in nitrogen. This is a fantastic liquid fertilizer that can be used on salad greens, spinach, cabbages and more. This will provide plants with an increase in Nitrogen levels and will allow them to develop sturdy and develop quickly and have deep, dark green leaves. The only thing you have to do is chop down nettles

allow them to soak, then spray the mixture on your plants.

* Seaweed: You could create your own seaweed-based liquid fertilizer by adding some seaweed in your water butt and let it sit in the water. It will be a fantastic fertilizer for all of your vegetable plants.

* Fish: Take some head and bones of a fish; place them into a container with an easy-to-fit lid. put the fish in a container and cover it with water. Put the lid on and shake it up at least once a week. Within 3 weeks to one month, the fish should be been broken down. Now you have high-quality fertilizer. It is rich in the trace minerals and elements the plants require. You should dilute this fertilizer liquid prior to applying it. You can put around 2 cups into the watering bottle and add water to it. This booster for plants will assist your plants reach their maximum potential.

* Manure: Place the manure into an hessian bag, and let it to sit in water before using it as a liquid food source for your plant.

Mulching

Mulching is the practice of placing some protective material on the surface of the soil, which is the area the area where you grow your veggies.

Mulching is utilized for numerous reasons. Mulching material will protect the roots of your plants from frost, the cold winter nights, or assist in stopping the loss of water and ensure that your veggies will not become dry.

There's a myriad of mulching materials you could make mulch from: straw, leaves and sawdust, hay or cardboard, grass clippings along with pine needles. There are some who even make use of carpet that is old, but there shouldn't be any carpet backing.

The ideal kind of mulch to use to use for your garden is dependent on your specific growing conditions. My preferred choice is grass cuttings.

Try out the various mulching options in order to discover what's most effectively for your needs. You are likely to find that mulching is influenced by your specific growing conditions as well as the kind of vegetable that you're cultivating.

A thing you must realize is that you should not do too much with these types of fertilizers. Don't make use of frequently or frequently. Utilizing these fertilizers for your plants' vegetable gardens is healthy, however should you make a habit of using them too often, they could cause harm to your plants since they may be exposed to an excessive amount of nutritional food. It is recommended to apply it at least once every two weeks or less.

Green Manure

Green Manure is one of the crops which is cultivated to enhance the soil. It's also a fantastic method to increase your garden's fertility using just an organic method.

It's an excellent method to enhance the soil structure and condition of your soil. If you're not making compost, the green manure can be the perfect way to boost the quality of your soil.

The only thing you require is an organic manure packet, a seeds, a fork and rake, and let the seeds fertilize stabilize, provide organic matter to the soil. The process is as simple as that.

There are many varieties of green manure options:

* You could plant quick growth green manures, such as phacelia, mustard and purple clover between your plants - these varieties of manure are able to grow rapidly between 4 and 12 weeks. What you must do is take them out or pull them to the ground.

Mustard is a great plant to plant in a particular area of the garden when you don't intend to plant or sow immediately - because it helps keep the soil free of weeds, and also give the soil some nutrients.

* You may also plant green manure, which is known as green manures for long-term like alfalfa, red clover and.

There are some who sow green manure, such as white clover in between slow-growing plants like cabbages, sweet-corn, or sprouts. After the plant is done, you can plant the manure in the soil.

If you're trying for ways to boost your soil's health, invest in some seeds of green manure to start right away. Your soil will appreciate it for your efforts and provide your taste buds with more delicious veggies.

Chapter 14: Tools You Need

If you're gardening in a small space, there aren't many equipment. Therefore, the equipment are required for starting will be fairly simple.

It is possible to purchase other equipment as you'll need these.

The tools are available in a variety of lengths and sizes nowadays, and you'll be able to find tools just right to your requirements. You'll need:

* Spade - Choose one that has the handle extending close to your waist, which makes it much easier to hold. It is possible to purchase very specific spades that are designed for various types of soil. A standard spade can be adequate for most soils.

* Fork: As the one above, be sure that it is placed at your waist for it easy to use.

* Rake can be very beneficial for a good way of arranging things in the garden.

* Hoe - there exist two kinds of hoes: the digging or cutting type as well as the drawing hoe. However, there are a variety of types and sizes of hoes that are available for the gardener. Choose one that feels right to you.

* Trowel.

* Hand cultivator for small hands.

* Secateurs: Try and get a good pair that will last for years if you take care of them. They are great for trimming items back.

* Gardening gloves: I love waterproof gloves which are also breathable, to ensure my hands aren't overly sweaty. Also, I prefer to wear gloves to prune roses and other prickly plants and ensure that your hands remain unprickly.

* Dippers can be very useful when making plant beds.

* Mat: a mat that you can kneel upon in case you have to kneel for a certain garden tasks.

* Garden Line - - you can create this by your own. Take 2 sticks and some long string with a strong length that is the same as the plot of garden. This will ensure you're sowing or planting in an even direction.

A wheelbarrow is perfect for the garden. It is able to carry heavier weights and heavier loads, unlike if you needed to transport it. There are numerous kinds of wheelbarrows, so you can choose one that meets your needs.

Tools can be bought at your neighborhood garden store or purchase the tools on line.

Most important is that you purchase a high-quality garden tools as they last longer and will have less of a chance to fail.

As you advance in your gardening experience You will realize that you will require specific equipment to accomplish certain tasks and you will be able to purchase. Do not spend money on tools initially that will never be used.

Looking after your tools

Create space within your shed to create a space for every tool. You can also put them up. This makes it easier to locate them.

It is important to take care of your tools. Clean off the dirt after a day of gardening. Rub the metal parts of the tool using a bit of grease to prevent corrosion from developing.

Additionally, if you keep the wooden tools you use outdoors, the wood will start to decay in time.

Additionally, if you keep your plastic-handled tools outdoors, it is likely that they'll be brittle and snap in a flash.

Weeding your garden

It's the easiest way to remove the weeds while they're tiny and small. You could use a hoe, or remove any weeds you encounter manually.

You should weed your garden regularly in order to stay at the top of your game If you let the plants to bloom, you could result in problems in your garden. Try to do weeding regularly, at minimum once per week.

Some weeds can be eaten. The most edible weeds are dandelions, nettles and chickweed. These weeds are all extremely healthy and tasty too. In particular, nettles are high in iron. They can be consumed as a vegetable, or even in soups.

Perhaps it is worth trying to keep weeds out of the corner of your vegetable area since weeds can be an excellent food source for insects, which are good for the gardening.

In the fight against bugs

If you own a vegetable garden, then you'll eventually encounter bugs that are causing trouble for your plants. There are a variety of natural methods for eliminating the majority of garden bugs.

As an example, slugs or snails are a major problem; the best way to get rid of them is to set certain beer traps. The best method is to cut a piece of grapefruit (that's consumed) dig a small hole in the soil and then place the grapefruit portion into the hole, making sure that it's at a level with soil. Fill the hole with beer.

* Ladybirds: This lovely insect is known to eat mites and aphids. Ladybirds can be attracted to your yard by planting blooms they enjoy including daisies, yarrow or even apricots. You

could also purchase ladybird larvae online However, I'd recommend you place some blooms in your garden to lure the birds to remain.

Enter "Buy Ladybirds" into your search engine and you'll be able to find an online retailer.

* Lacewings: They devour Aphids, with the additional bonus that their larvae devour Aphids.

The beneficial bugs eating insects could be drawn to your in your garden by planting different flowering plants such as yarrow. They also need to dedicate only a tiny area for wild plants to draw this insect. It allows you to attract naturally-occurring predators to your yard.

It is also possible to use natural methods for keeping your garden safe from pests. These solutions are completely safe and natural.

Use a spray bottle to fill it with water. Add 1 tablespoon of liquid soap. shales and then spray both on the upper and lower leaf of vegetables which have mites or aphids on the leaves. This kills them.

A naturally occurring substance known as milky spore may be found all over the garden, and is waiting for grubs of beetles that could infect them with the deadly disease which will kill them prior to turning into beetles. Since this spore spreads throughout the gardens and believed to be effective for 40 to 50 years, it's something which you're only likely to need to perform one time.

Another option to consider is wormwood, which can be utilized as an all-purpose pesticide as well for sprays to repel snails and slugs.

Fungal illnesses can be treated using a solution made of baking soda mixed with water.

Although they're completely natural methods of killing bugs, it is best to only apply the sprays only if you absolutely need to, as the sprays that are natural could still kill beneficial bugs, too.

Seeds to choose from

There is a way to buy plants from the local garden center or nursery, and many supermarkets carry them in the season of spring. It is also possible to purchase seeds on the internet - there are numerous varieties of seed companies to shop with. Below are a few that I like:

You can do a an internet search on Google and you'll see a lot of other people in your area.

Most of the time, online seed firms offer a wider selection of seeds for you to select from.

Which are the best vegetables for you to pick?

As you begin your garden for vegetables You want to be able to observe the results fast to inspire your garden to continue growing.

Below are a few of the best crops to plant as you begin your journey.

Radishes: Radishes is one of the simplest vegetables to begin growing. They are fast growing and available in just weeks. The plants are also free of any diseases or pests.

Courgettes (zucchini) It is yet another

A vegetable that is easy to cultivate. You can

You can also purchase plants at your local garden store or buy some

center or sow 2-3 seeds into small pots

covered with compost, then be sure to cover them with a little

of of. It is recommended to begin these inside

They will require up to 10 days for them to take to pop

through. After that, you can the hardening process (see the previous paragraph)

Before planting them out. They'll reward you

You will be able to provide you with a variety of varieties of courgettes. If you don't

Choose regularly, the small courgettes are destined to grow into large marrows. It is possible to purchase the standard courgettes but you also can get the round form. Green is the most common colour but you can also find yellow courgettes, similar to the ones the above courgettes that we grow in our backyard. TIP: Water courgettes close to the base of the plant instead of around the leaves to keep away leaf mildew.

Swiss Chard: Swiss chard is like spinach or perpetual spinach. Swiss Chard may be entirely white, but it is offered in a rainbow of colors. It is easy to cultivate and beautiful to display at your table.

Lettuce is easy to grow. Pick a variety of loose leaves that means that you can choose the leaves whenever you'd like to. You can also pick a type of lettuce that you enjoy There are plenty choices to pick from.

Beetroot: Beetroot is an easily grown vegetable. Beetroot doesn't have to worry about any kind of pest or disease.

Tomatoes: They can be grown in all climates. They are among the most simple to grow.

It is possible to grow tomato seeds from seed, but a lot of individuals prefer to purchase them at the community garden store, especially when you just want to purchase some plants. You can get diverse kinds. Then you can plant them in the indoor or outdoor areas. The tomato needs support. You can put them on a string hold them up or give them a cane. It is also important to cut off the shouts from behind.

Strawberries are a simple plant to cultivate. The process of planting strawberries is simple. They can be planted inside pots, in special pots for strawberries or even in a the garden. Create a trench, then put the strawberry in and keep it watered. The mulch can be placed around the plant with straw, leaves or pine needles.

Chapter 15: Companion Planting

Companion plants are those that cultivate vegetables alongside other plants, and are mutually beneficial. The benefits of companion planting are immense when you're a gardener. It is particularly beneficial when you are looking to cultivate organically.

Many of them emit smells that can attract good bugs or deter insects that are harmful. It's an excellent method that is in harmony with the natural world and can help grow strong and abundant crop without using sprays or pesticides.

The companion planting method is great to keep garden pests out. Through combining plants, they aid each other in to keep pests out - they accomplish this by drawing beneficial insects. i.e ladybirds.

Companion Planting Combinations

Plant marigolds near to tomatoes. Since marigolds emit a strong smell that repels blackflies, greenflies and aphids. However, marigolds can also draw hoverflies, which are fond of devouring Aphids.

* Plant nasturtium in the vicinity of the cabbages. Because nasturtiums can be beloved by butterflies of all colors. They can lay their eggs the ground and let your cabbages to themselves.

* Planting leeks or garlic in close proximity to carrots may confuse insects that feed on carrots and guard against damage from slugs. (Have noticed before that garlic does not suffer the slug damage?)

* Basil is a great plant to keep away whiteflies when planted with plants.

* Nettles - loved by white butterflies of cabbage which keeps them from the brassica plants.

Make sure to plant your other plants in the same time with the vegetable crops to prevent insects from getting an early to get a head.

A - Z of Vegetables

There are so many varieties of vegetable seeds that gardeners may not be aware of. My parents operate a garden at the market, they would plant every type of vegetable that was possible. Certain of the veggies that are listed here are cultivated by a variety of gardeners, and there is a few that aren't or not known to most vegetable gardeners.

I'm listing the veggies according to alphabetical order.

* Artichoke - Globe. They grow in large plants and the globe grows above the plant.

* Artichoke - Jerusalem. They grow in the ground and remove them as you would potatoes. They're notorious for growing and taking over!

* Asparagus - Are a delicious vegetable. It is possible to grow them on your own however you must be cautious. If you purchase asparagus crowns for planting, they will need to be waited for three years before you are able to correctly pick the crowns. The asparagus crop will last over the course of twenty years.

* Asparagus Pea - It's more than a "mange tout" it is best to pick the pots at approximately 1" length, otherwise they will become hard. It's a plant that produces an exquisite red flower.

* Aubergine - Is a large purple fruit. It is best to grow them in a greenhouse, but you could try growing the fruit in a protected area.

* Basella - Malabar Spinach. It is a climber spinach that grows about 3 feet tall and must be grown inside a greenhouse or poly-tunnel. It is a delicious spinach and tastes a little milder that regular spinach. The leaves have a deep green.

* Beetroot - Many gardeners plant beetroot in their garden. Beetroot is an amazing veggie that a lot of people like.

* Borecole or Kale. An excellent plant for when you are unable to find leaves in winter when there isn't a lot of vegetables available.

* Broad Beans are a plant which you can either like or dislike.

* Broccoli There are a variety of kinds available.

* Brussel sprouts. We really love brussel sprouts.

* Cabbage - Various varieties like white, red, green and the savoy. There's a variety you'll like to cook and cultivate.

* Cape Gooseberries - Physalis. Actually, it's the fruit. However, they were in my parents' list. Cape gooseberry is a tasty tiny fruit about which is about the size of a gooseberry as well as the bright color of orange. The husk is a papery material surrounding their berries. It is best to plant them in a greenhouse, but can thrive in a sunny, sheltered area. It will grow up to 6 feet.

* Cardoon is a real vintage vegetable. It's similar to an artichoke. It is edible.

* Carrots-We all grow the old fashioned carrots.

* Cauliflower - Different types that include purple, white or.

* Celery - green and red, or blanched.

* Celeriac - It is a round, spongy ball similar to the turnip or swede. It is similar to celery.

* Chinese cabbage - Could be consumed in the same way as lettuce.

* Chicory - - Chicory is much more well-known in Europe than in the UK. The plant is well worth the effort to grow.

* Courgettes-We have grown yellow and green courgettes as well as courgettes that are round.

* Cucumbers - Can be outdoor or indoor cucumbers.

* Endive is similar to the lettuce, but it is more bitter.

* French Beans - Many different kinds: purple, green or yellow.

* Garlic comes from the family of onions. Garlic is delicious fresh, or let it dry out and remain for the entire the winter.

* Gherkin * Gherkin Grow Gherkins if you have them. You must pick the gherkins!

* Kohl Rabi. I love turningips and can cultivate a green or purple type.

*Leeks belong to the part of the onion family. Leeks can take quite a while to develop and can be an excellent vegetable during the winter.

* Lettuce - many different varieties: Iceberg, cos, butterhead, Lollo rosso, etc.

* Marrow- You can purchase specialized Marrow seeds to grow into marrows, or you'll discover the courgette you forgot to harvest and then it develops into an Marrow.

Melons are a good example of a fruit that can be grown indoors or outside. They will require a shaded space if they are grown outside.

* Okra - Ladies Finger. Origins are from Africa. It is best to grow it in a climate of the greenhouse.

* Onions - Onions generally grown by gardeners of all kinds.

* Parsley * Parsley Parsley is an herb that is a great plant to grow in your backyard.

* Peas, which includes the mange all.

* Peppers: Green Red and Yellow.

* Peppers Hot chilli.

* Perpetual Spinach is a type of spinach that tends to last longer than regular spinach.

* Potatoes: We only had first and later potatoes however, you could also plant maincrop potatoes.

* Purslane - Golden or Green. It is eaten as young leaves.

* Rhubarb * Rhubarb Rhubarb is quite simple to cultivate and produces lots of rhubarb.

*Runner Beans need an environment that's safe for them for them to flourish, then they'll grow a great number of beans to help you.

* Salsify - - Salsify is the largest tapered root, with a soft texture. The root is a great addition to stews for extra the appearance of a stew.

* Scorzonera * Scorzonera Scorzonera is a texture looks like parsnips and has a flavor similar to artichokes.

* Seakale: You can enjoy this as asparagus. You can blanch it or have it eaten in its green.

* Swede is a swede is a hefty root vegetable that can be utilized in stews, and can also be used to make Cornish Pasties.

* Sweetcorn is a tall plant which produces three to four cobs every plant. Sweetcorn grown at home tastes more delicious than store bought.

* Swiss Chard Swiss Chard, like spinach. It is white and has stunning green leaves. It are able to get it in rainbow colors.

* Turnip - They are a root veggie and can be enjoyed alone along with butter, the addition of salt or pepper.

* Tomatoes: There are various kinds of tomatoes. They can be dressed outdoors while many are cultivable inside. Nothing is

better than the tomato you've created by yourself.

Some Common Gardening Terms

Here's a listing of terms and words used in gardening as well as their meanings.

* Drill is a small V-shaped trench that you can use to plant seeds into. It is possible to use a hoe, the edge of a rake or even the edge of a trowel to construct the drill.

* Germinate is once the seed has been planted and begins to develop and form a root and stem.

* Harden off is the term used to describe that you need to adjust new plants, that are grown indoors, outdoors, to the weather outside. It is necessary to move your plants outdoors at times during the day and bring them back inside in the evening over a period of time.

* Prick out, which refers to taking tiny plants that are 1 or 2" and place them in larger pots so they develop. It is used for tomatoes.

Seeds are planted after which the plants are then put in.

After sowing, you can thin and after the seeds are sprouting, the seeds need space, so you need to can thin them. There are times when you can make use of thinned plants and then put them elsewhere to grow.

Transplants and seedlings are plants that are still young.

* Transplant: move the young plants from one growth location to another.

* Under cover means an element of protection in order to shield plants or seeds from cold or bad weather. I.E. cold frames, cloches greenhouses, tunnels.

Chapter 16: Your First Garden? Getting Started.

Fresh vegetables, fresh herbs All organically grown ... Who would not want to be a part of it? It's a sure thing that you would love it don't you? Since organic is a new trend across the globe right now. If you want to go organic in your food choices, then you should be able to grow the food yourself. It is impossible to know if what you buy at the store can be truly natural.

Beginning the first garden can be difficult or simple according to the method you take. However, I'll give you one tip: to ensure that it is easy begin small!

The idea of a large backyard for a beginner gardener will not be a great idea as it could make you feel frustrated. It is best to begin by making it small, and you will be able to extend it after you've gained knowledge. There are many aspects associated with gardening, particularly when you're a beginner.

DECIDE ON THE SPACE

It's the initial step. What exactly do you want your garden to look? Outside the home? in front? You must find a spot with a shaded area, where your flowers, plants and veggies will be safe from the elements, but in the same way, they will receive enough sunlight.

Finding a suitable space to grow can be more difficult than you believe. Many vegetables, like cucumber, elephant garlic tomatoes, quinoa Jerusalem artichoke, tomatoes and vines require around six hours of sunlight. Therefore, it is important to be sure that the location where that you plant your garden is getting enough sunlight.

There is a need for shade, too. Certain crops require greater shade than sun. Therefore, you should look for the area to get sun as well as shade during different time of daytime.

PREPARING THE SOIL

This is where the actual work starts. Prepare to get involved. The work isn't only about turning and tilling the soil around It is concerned with knowing the pH balance and acidity and alkalinity in the soil.

Be aware that it's crucial to match the proper plants to the appropriate soil pH. As a general rule that you must be aware of is that the range to measure alkalinity, or acidity, of soil is from the range of 0-14. That means soil that has an acidity of 8.0 is considered alkaline, whereas soil that is pH six or lower is acidic. If the pH falls below 7.0 the soil has a neutral pH (neither acidic or alkaline).

Certain crops, like tomatoes, do very well in soils that are acidic, such as this, for instance. It requires a pH value that is at or below 5.5. Eggplants require a pH that is 5.5 to 6.5 as well as cucumber requires a pH range of 5.5 to 7.0 and garlic demands soil that has a pH between 5.5 up to 7.5.

A few plants also require soil that is alkaline than the natural. For instance, artichoke needs that it be planted in soil that is pH range that is 6.5 or 7.5, Jerusalem artichoke needs 6.7 to 7.0 pH while Tomatillo demands an acidity level between 6.7 up to 7.3 and and on.

In the end, it's clear that you must determine the pH levels of the soil where you plan to plant your plants. This way, you'll know if you need to make it more acidic or more alkaline. It may be necessary seeking out professional help But don't be concerned. It will be worth it if you harvest a great crop.

But, you might also need to learn what you can do to determine the soil's pH for yourself. It's not a bad idea to be aware of this information, doesn't it today? It's quite simple actually. All you require is vinegar. You can pour one-half cupful of vinegar in the soil, and observe what happens. If the soil is acidic the vinegar will bubble and when it's acidic and not acidic, it won't.

There is a straightforward method to measure the acidity, or alkalinity of soil. It is necessary to use a pH test kit, which can be found in local stores for gardening. Simply dig a tiny space in your garden and fill it up with water that has been distilled.

By using your test probe place it inside the hole, and then hold it for several minutes. Take the measurements from the scale. If it reads 7 or less, then the soil has neutral. If it's above or less than 7, your soil may be alkaline or acidic.

TO DIG OR NOT? DIG!

Don't let yourself be unproductive! gardening is hard work However, the result of fresh produce and the herbs are far more worth the effort. It is therefore more beneficial to dig, as:

* This will help make the soil looser, making it much easier for roots be able to

* This will allow for enough water permeation

It is possible to make the soil level following digging

It is simple to distribute organic matter over the soil, to enhance the quality of your soil.

* This will allow you to plant cuttings, seeds or seedlings more simple.

Now, you'll be able to be able to see how simple it is to have your garden space prepared for planting your beautiful quinoa Jerusalem artichokes, eggplants as well as squash, among others. The preparation of the garden is one of the main aspects of gardening.

Chapter 17: Preparing A Raised Bed

If you're a novice with gardening, it's essential to know the various kinds of beds are suitable for vegetables. They can be raised, flat and sunken.

We will take a discuss the process of preparing the raised bed. The name says it all it can be found a few inches above the ground, usually about at a height of 11 inches. The beds can be used for vegetables and fruits (for instance, cucumbers as well as the elephant's garlic courgette and spinach).

In the beginning, you must clean the area on which you will create your raised bed similar to what we have discussed in Chapter 1.

Next is to dig up. This is helpful in loosening soil that improves drainage, and facilitates root growth to penetrate. Because a garden usually only a tiny piece of land, it's essential to inform you. Beware of machines since they can impact the soil's foundation.

Also, you can enhance the soil. In particular, clay soil isn't ideal for growing veggies like tomatoes and garlic. This is why you should incorporate organic manure into the soil's drainage as well as root penetration.

When the bed is tilled then it's the time to add soil so that you reach the height you want to reach. One thing to keep in mind is that it is possible to change the soil type within the bed in case it doesn't fit with the crop you are growing. Transfer soil from an area for use in the raised bed. You can till the soil in order so that it blends well.

The final thing to do is add organic manure into the raised bed. It is beneficial because manure is organic and doesn't contain any synthetic chemical. Be sure to select high quality manure that is well-prepared. It is possible to make yourself at your home. Manure that is not properly prepared can affect the soil and may scorch or even burn your veggies.

When you're done after you have completed your bed, it is crucial to make a border around your bed in order to ensure it is in a good form, provide the necessary support, and also to repel rodents and pests. It is possible to use bits of timber or bricks for a border around the bed.

If you follow the guidelines earlier, remember that the bed's size is important. be sure that you are able to access the bed from all directions to avoid entering it. Make sure you have the size that is 1.5 meters.

Your footprints can definitely impact the soil, as pressure creates pressure. Additionally, your steps will not appear attractive since they are asymmetrical upon the soil's surface.

Chapter 18: Choosing High Quality Seeds

If you're thinking of laying an area for your garden with tomatoes, spinach, or other fruit you're thinking of It is crucial to think about purchasing high quality seeds. It is the type of seeds that you consider when you decide if you'll be eating healthy veggies or otherwise.

What to look in selecting the best seeds are:

1. What's the chance of developing germs?

The main thing to consider is the likelihood of the seed being able to develop the needed elements in an agricultural plant. Take a look at the data in the tag of your seed to determine the percentage of germination.

A high-quality seed will have an germination rate of 80 percent and over. It means the chances of them germinating are more likely than being unable to germinate.

2. What is the best way to use it to grow above ground?

When you've verified the seed's germination, you must to determine how the seed is growing above soil level. This will allow you to know the best way to set your bed. For instance, if you want to plant tomatoes, seeds usually are very tiny. So, it is important to create a fine soil and your holes to plant are shallow. The seeded tomato won't be grow in a hole that is very large and deep.

3. You can only get the purest varieties of seeds

Be sure the seeds you choose contain high levels of quality. Do not grow vegetables such as Jerusalem artichoke or vines, and then harvest various varieties of the same plant.

It is important for each of seeds to be uniform in high-quality. Check out the details on the seed bag and, most importantly, inquire, conduct some online research to find out who has the finest quality.

4. Growth rate

This is related to the speed at which the vegetable develops. If, for instance, you plant spinach, how rapidly will it grow? Thus, you must select seeds that will produce healthy vegetables.

5. Get yourself a attractive people

It is in line with the way in which the particular kind of plant grows, as that affects the appearance of it. The look of your vegetables are crucial, particularly in the case of growing vegetables for sale. If they look unhealthy it is possible that appetites will be affected. If it's attractive that means it is healthy to eat. Well most of the time it is.

6. Do they have the ability to fight the spread of

The best seeds can help to reduce costs when treating diseases. So, make sure that the seeds you buy will endure illnesses.

7. Seek out seeds with the lowest amount of inert matter

The bag of seeds isn't as pure as they wish to convince you however, you must ensure that you buy an item with as few stones, sticks, leaves and other impurities as is possible.

8. Only purchase seeds that are certified.

Be sure to purchase your seeds from manufacturers or suppliers which have been endorsed by several reputable organizations. This is among ways to stay away from buying inferior seeds of low quality offered by fraudulent sellers on the marketplace.

Chapter 19: Planting Your Vegetables

Spacing

If you are planning to plant your veggies spacing is the most important element to be considered. The requirements for spacing of different kinds of vegetables vary such as the space required for leafy vegetables is different from the non-leafy types. It is therefore essential to make sure that the accurate spacing measures follow.

The proper spacing is designed to maximize your yields from your garden while simultaneously maintaining the health of your crop.

Two types of space that you should keep in your mind:

* The spacing between rows is determined by the row space. the rows

* Crop spacing is the distance between plants

As an example, the recommended space for Jerusalem artichokes ranges from 12 to 18 inches. The recommended row spacing ranges from 24 to 36 inches. Cucumbers must be planted between 12 inches between plants as well as 18-72 inches in between the rows.

The recommended spacing of the plant is 2 to 4 inches, and row spacing is 12-24 inches. There's a huge variation between the plant and row spacing for spinach. It is due to the fact that it is the leafy plant.

The tomatoes should be placed at 18 to 36 inches in between the plants and between 24 and 48 inches in between rows. The cucumbers should be planted at 12-18 inches between the individual plants.

A different way to measure spacing is to think about how big the plant grows when it expands. This way, for plants like cucumbers, melons and even pumpkins that are in a spread-out manner in rows, spacing between the rows should be spaced at 72 inches.

For plants that are small, like leeks and leaf lettuce it is recommended to space them four inches of space between the plants.

How do you ensure that you place your veggies in a proper way?

1. This reduces the spreading of disease and pests between plants and another, as the plants won't be near each other.

2. It decreases need for lighting, water, as well as nutrients. In the end, crops with a tight spacing can result in becoming overcrowded, causing a scramble to get the nutrients available as well as light. The result could be crop varieties becoming unhealthy or slow in growth and, consequently, the yields will be low.

3. It helps the plants increase in size, and especially the leafy veggies

4. There is plenty of room to do your work - you don't want to harm the vegetables you eat due to how you move in sleeping area.

5. Have easy access to the garden. What is the purpose of having an area that isn't going to be easily accessible? It is not your intention to create a situation that makes harvesting,

adding manure or weeding a nightmare are you?

6. This prevents wasteful use of garden space. They are neat, and extra space could be utilized for different landscaping demands.

7. Space is the only way to ensure that all parts of a plant are covered with pesticides

The spacing is also dependent on what part of the crop that you are planning to harvest. Are you going to harvest the root or just the leaves? In the case of a root vegetable like Jerusalem artichoke will require sufficient space on the ground while the leaves or top aren't a huge difference.

Tomatoes, for instance, and spinach require enough "leaf space' since the harvest is contained in the leaves.

Chapter 20: Mulch And Green Manure For Your Vegetable Garden

An organic vegetable garden provides fresh, healthy produce. The produce is not modified or altered in any way. There is no way to know for sure everything you purchase at supermarket is organic. As a matter of fact, you may not need organic fertilizers. This is the reason why the green manure and the mulch comes into.

Green manure

Like the name implies the substance is organic and green manure. It's a form of manure in which you are able to grow certain varieties of vegetables to your backyard. If they're still fresh, you will need to remove them from the garden and then plug in the ground. These plants will decompose slowly and will make the garden more suitable for growing your crops. Simple as that.

Manure that is green is planted immediately after harvesting to ensure that the following crop you plant will benefit from the nutrients.

Specific characteristics of the plants utilized as manure in the green form:

* Growth that is vigorous

* They're leaving

* Injects nitrogen back into the soil (plants that have root nodules)

* Adjusting your local climate

* It is considered a cover crop (one that is spreads across the surface)

Some examples of plants that make green manure would be soybeans as well as sweet clover.

Green manure has many benefits.

* Enhances soil fertility

* Helps improve the structure of soil

* Ensures that the soil is properly aerated. the soil

* It wards off weeds by slowing the growth of these plants

* It enriches soil with nitrogen. This occurs with the beginning of the season of rain. This is why green manure aids greatly when you plant in dry seasons, where the crops are dependent on irrigation

Mulch

Mulch can be described as any type of material used to cover soil after your plants are sprouting. Mulch may be either organic (for instance straw) or even inorganic (such as plastic or polyethylene material). Organic materials are made of living matter, such as plants. If you're looking to create a natural, healthy garden, choose organic mulch, such as straw. It is beneficial if are unable to get organic mulch.

The importance of mulching your garden

* Make sure there is an adequate retention of moisture in the soil.

* Helps prevent the loss (evaporation) of the nutrients in the soil

* Improves the fertility of soil. It is used for the mulch made of organic material.

* Weed growth is decreased.

* Controls the temperature of soil. mulch is a shield from scorching sun and keeps the warmth of the soil during evening. This is essential for living microorganisms that live in the soil which promote soil fertility.

Mulching helps prevent erosion since it decreases the velocity of the water that is on top of soil.

* Controls pests.

Also, it provides an aesthetic value to the garden.

It is evident that by composting manure can provide numerous advantages. Manure not only improves the quality of the soil by supplying it with vital and essential nutrients, but it can get more yield. The fact is that nutrients without water won't do your crop much better. So, mulching is a good way for better moisture retention.

Chapter 21: The Art Of Growing Up Rather Than Growing Out

Growing older or getting bigger? It's the question. It is not the only question. Shakespeare made a comment in Hamlet to be, or not ...

Constructions rise rather than down. The people would prefer to develop "up" than outwards. So why not plant vegetables?

This is one of the questions which a majority of gardeners face when deciding which approach to employ. Plants that grow upwards are known as vertical gardening. Growing tall is the best method of making the most of your garden space. The reason is that those with smaller gardens need to cultivate the most vegetables they can using a vertical support.

Trellising

Large gardens are those who be inclined to lend physically to the vegetables they grow. A few of the veggies such as tomatoes and strawberries have a weak point to varying degrees. It is important to offer the necessary support when they begin to grow. It is referred to as Trellising. Other fruits and vegetables that require trellising are cucumbers, and tomatoes.

It is possible to give less affluent vegetables some support by placing them on the fence around your garden. But, it is important to make sure that the plants grown on your fence are able to be rotated through the seasons.

Utilize the fence

This fence is therefore performing the double job and is an excellent method of using the fence to its fullest. Another method of trellising is built using metal or wood. In order to ensure the effectiveness of your vegetable support ensure that they are set up prior to the time your plant needs its support, or prior to planting the plant.

In the case of certain veggies, like tomatoes, it is possible to secure the plant to a support, or the greatest care weave them around the support when they begin to expand.

Pin them to tomatoes

Another method of making sure the growth of your plants rather than out. This is especially important with tomatoes because the plant prefers to expand rather than out. This creates the creation of a smaller and more tightly controlled plant, which can yield greater yields. This can affect the shape and size of the entire plant. It also helps new growth emerge more quickly.

You can grow more within a smaller size

Being able to grow rather than out will yield a more favorable result in terms of reducing the available space and by paying attention to the trellis because they're less vulnerable to plant diseases and insects.

The process of watering plants becomes simpler. So, it's recommended to plant the plants in rather than spread out.

Chapter 22: How To Care For And Feed Your Vegetables

It is not something that is easy to do however, it's easy once you have a basic understanding of facts about gardening. It's ok, but the result, that plethora of organic and healthy veggies is surely much more worth the time and effort spent.

Inattention to your vegetable garden will cause death the plants and you'll waste your time and effort making the seeds. The garden is planted the seeds and your plants have begun to sprout. This is what you'll maintain your garden vegetable.

Regularly drinking

Regularly watering your plants is the best method of maintaining your plants. If you don't give them water, they'll start to die and will expand at a snail's pace and then die. Keep the soil dry but not wet. Sprinklers, furrows, or drip systems to water your plants.

Coverage for the Mulching

Mulching is a different method of taking care of your vegetable garden. It involves the application of organic matter to increase the water retention in the soil as well as regulate temperature in the soil. It helps minimize plant diseases, and keep the soil from being eroded. It ensures your plants are protected from diseases, and the water of the soil is preserved.

Sprinkle some fertilizer on top.

Fertilizers are yet another method to care to your garden. Many vegetables require that you apply fertilizer prior to planting which will last throughout the season.

For instance, some vegetables, like tomatoes might require fertilizer on a regular basis. So, your veggies are wholesome and yield will also increase.

Make sure you are weeding properly - decrease the number of competitors

They are the gardener's most dreadful enemies as they fight the plants for light, water and food. They also harbor insects and diseases. Eliminating weeds on a regular basis will make sure that veggies get adequate nutrients to sustain their development. This can be done by pulling or cultivating by hand.

Pest eradication

Pests could harm your vegetable plants and spread diseases. Therefore, you should inspect your crops regularly for insects and take action to stop this threat immediately.

Spraying plants with insecticides is the best way to go about stopping the pests and eliminating the pests. It is recommended that the insecticide be suitable for eating plants so that carcinogens are not a problem.

Make sure you get rid of any decaying or dead plants as they can become an ideal habitat for insects as well as diseases.

Not least of all, make sure you harvest your vegetables when they are ready, at the right time, not when they're fully cooked or long after they're scheduled to be harvested.

www.ingramcontent.com/pod-product-compliance
Lightning Source LLC
Chambersburg PA
CBHW071442080526
44587CB00014B/1963